TRIGONOMETRY

Maths Solutions

50 CHALLENGING MATH PROBLEMS WITH SOLUTIONS

Blank page

Contents

Blank page

Useful Formulas

Addition Formulas

- $\sin(x + y) = \sin x \cos y + \cos x \sin y$

- $\sin(x - y) = \sin x \cos y - \cos x \sin y$

- $\cos(x + y) = \cos x \cos y - \sin x \sin y$

- $\cos(x - y) = \cos x \cos y + \sin x \sin y$

- $\tan(x + y) = \dfrac{\tan x + \tan y}{1 - \tan x \tan y}$

- $\tan(x - y) = \dfrac{\tan x - \tan y}{1 + \tan x \tan y}$

- $\sin 2x = 2 \sin x \cos x = \dfrac{2 \tan x}{1 + \tan^2 x}$

- $\cos 2x = \cos^2 x - \sin^2 x = \dfrac{1 - \tan^2 x}{1 + \tan^2 x}$

- $\tan 2x = \dfrac{2 \tan x}{1 - \tan^2 x}$

- $\sin 3x = 3 \sin x - 4 \sin^3 x$

- $\cos 3x = 4 \cos^3 x - 3 \cos x$

- $\tan 3x = \dfrac{3 \tan x - \tan^3 x}{1 - 3 \tan^2 x}$

Factorization

- $\sin x + \sin y = 2 \sin\left(\dfrac{x + y}{2}\right) \cos\left(\dfrac{x - y}{2}\right)$

- $\sin x - \sin y = 2 \cos\left(\dfrac{x + y}{2}\right) \sin\left(\dfrac{x - y}{2}\right)$

- $\cos x + \cos y = 2 \cos\left(\dfrac{x + y}{2}\right) \cos\left(\dfrac{x - y}{2}\right)$

- $\cos x - \cos y = -2 \sin\left(\dfrac{x + y}{2}\right) \sin\left(\dfrac{x - y}{2}\right)$

Defactorization

- $\sin x \sin y = \dfrac{\cos(x-y) - \cos(x+y)}{2}$

- $\cos x \cos y = \dfrac{\cos(x-y) + \cos(x+y)}{2}$

- $\sin x \cos y = \dfrac{\sin(x+y) + \sin(x-y)}{2}$

Trigonometric Equations

- $\sin x = \sin y \quad \Rightarrow \quad x = n\pi + (-1)^n y \quad Where \quad y \in \left[-\dfrac{\pi}{2}, \dfrac{\pi}{2}\right]$

- $\cos x = \cos y \quad \Rightarrow \quad x = 2n\pi \pm y \quad Where \quad y \in [0, \pi]$

- $\tan x = \tan y \quad \Rightarrow \quad x = n\pi + y \quad Where \quad y \in \left[-\dfrac{\pi}{2}, \dfrac{\pi}{2}\right]$

Trigonometric Identities

- $\sin^2 x + \cos^2 x = 1$

- $1 + \tan^2 x = \sec^2 x$

- $1 + \cot^2 x = \csc^2 x$

- $\csc x = \dfrac{1}{\sin x}$

- $\sec x = \dfrac{1}{\cos x}$

- $\cot x = \dfrac{1}{\tan x}$

Inverse Trigonometric Functions

■ $\sin^{-1} x = \csc^{-1}\left(\dfrac{1}{x}\right)$

■ $\cos^{-1} x = \sec^{-1}\left(\dfrac{1}{x}\right)$

■ $\tan^{-1} x = \cot^{-1}\left(\dfrac{1}{x}\right)$

■ $\sin^{-1} x + \csc^{-1} x = \dfrac{\pi}{2}$

■ $\cos^{-1} x + \sec^{-1} x = \dfrac{\pi}{2}$

■ $\tan^{-1} x + \cot^{-1} x = \dfrac{\pi}{2}$

$\sin \sin^{-1} x = x$	$\sin \cos^{-1} x = \sqrt{1 - x^2}$	$\sin \tan^{-1} x = \dfrac{x}{\sqrt{1 + x^2}}$
$\cos \sin^{-1} x = \sqrt{1 - x^2}$	$\cos \cos^{-1} x = x$	$\cos \tan^{-1} x = \dfrac{1}{\sqrt{1 + x^2}}$
$\tan \sin^{-1} x = \dfrac{x}{\sqrt{1 - x^2}}$	$\tan \cos^{-1} x = \dfrac{\sqrt{1 - x^2}}{x}$	$\tan \tan^{-1} x = x$

$\sin^{-1}(-x) = -\sin^{-1} x$	$\cos^{-1}(-x) = \pi - \cos^{-1} x$	$\tan^{-1}(-x) = -\tan^{-1} x$
$\csc^{-1}(-x) = -\csc^{-1} x$	$\sec^{-1}(-x) = \pi - \sec^{-1} x$	$\cot^{-1}(-x) = \pi - \cot^{-1} x$

Properties of Triangle

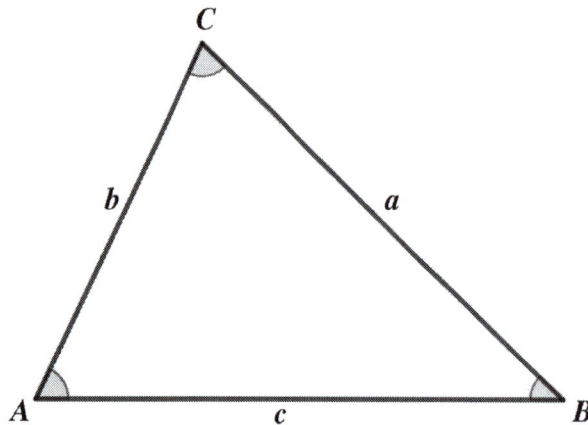

$$s = \frac{a + b + c}{2}$$

$\cos\left(\frac{A}{2}\right) = \sqrt{\frac{s(s-a)}{bc}}$	$\sin\left(\frac{A}{2}\right) = \sqrt{\frac{(s-b)(s-c)}{bc}}$	$\tan\left(\frac{A}{2}\right) = \sqrt{\frac{(s-b)(s-c)}{s(s-a)}}$
$\cos\left(\frac{B}{2}\right) = \sqrt{\frac{s(s-b)}{ac}}$	$\sin\left(\frac{B}{2}\right) = \sqrt{\frac{(s-a)(s-c)}{ac}}$	$\tan\left(\frac{B}{2}\right) = \sqrt{\frac{(s-a)(s-c)}{s(s-b)}}$
$\cos\left(\frac{C}{2}\right) = \sqrt{\frac{s(s-c)}{ab}}$	$\sin\left(\frac{C}{2}\right) = \sqrt{\frac{(s-a)(s-b)}{ab}}$	$\tan\left(\frac{C}{2}\right) = \sqrt{\frac{(s-a)(s-b)}{s(s-c)}}$

Projection Formula

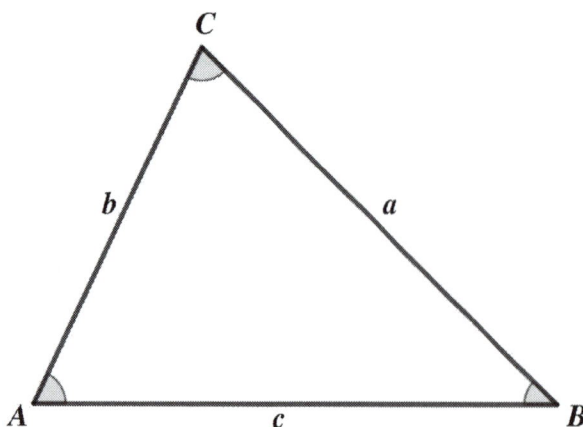

$$c = b\cos A + a\cos B$$

$$a = b\cos C + c\cos B$$

$$b = a\cos C + c\cos A$$

Cosine Rule

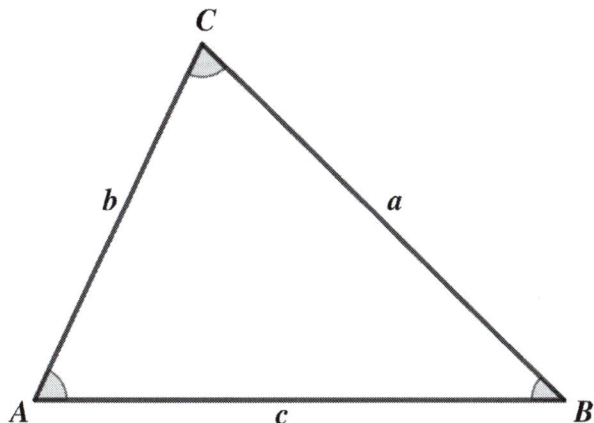

$$a^2 = b^2 + c^2 - 2bc\cos A$$

$$b^2 = a^2 + c^2 - 2ac\cos B$$

$$c^2 = a^2 + b^2 - 2ab\cos C$$

Sine Rule

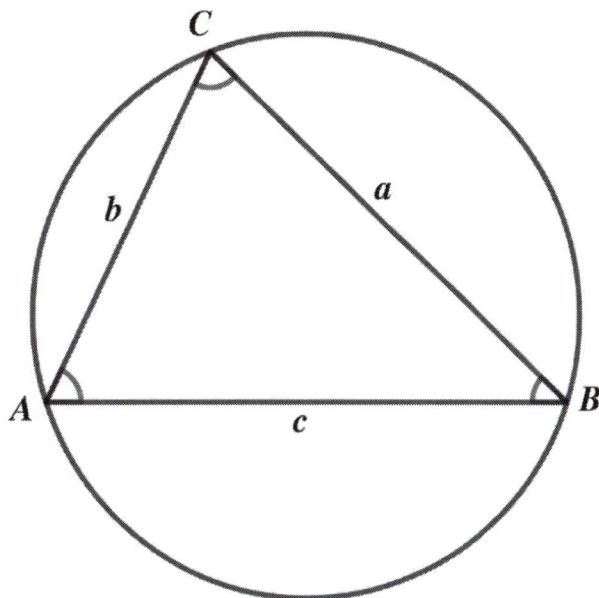

$$\frac{a}{\sin A} = \frac{b}{\sin B} = \frac{c}{\sin C} = Diameter\ of\ the\ circle$$

Trigonometric Functions

Problem 1

Find the exact values of

A. $\sin 18°$

B. $\cos 36°$

C. $\tan 15°$

Solution

A.

Let $\theta = 18$ then $90 = 5\theta$

$90 - 3\theta = 2\theta$

Taking sine both sides, we get

$\sin(90 - 3\theta) = \sin 2\theta$

$\Rightarrow \quad \cos 3\theta = \sin 2\theta$

$\Rightarrow \quad 4\cos^3 \theta - 3\cos \theta = 2 \cdot \sin \theta \cdot \cos \theta$

$\Rightarrow \quad \cos \theta \cdot (4\cos^2 \theta - 2 \cdot \sin \theta - 3) = 0$

$\Rightarrow \quad 4\cos^2 \theta - 2 \cdot \sin \theta - 3 = 0 \qquad (\because \ \cos \theta = \cos 18 \neq 0)$

$\Rightarrow \quad 4(1 - \sin^2 \theta) - 2 \cdot \sin \theta - 3 = 0$

$\Rightarrow \quad 4 - 4\sin^2 \theta - 2 \cdot \sin \theta - 3 = 0$

$\Rightarrow \quad 4\sin^2 \theta + 2 \cdot \sin \theta - 1 = 0$

$\Rightarrow \quad \sin \theta = \dfrac{-2 \pm \sqrt{4 - 4 \times 4 \times (-1)}}{2 \times 4}$

$\Rightarrow \quad \sin \theta = \dfrac{-2 \pm \sqrt{4 + 16}}{8}$

$\Rightarrow \quad \sin \theta = \dfrac{-2 \pm 2\sqrt{5}}{8} = \dfrac{-1 \pm \sqrt{5}}{4}$

$$\Rightarrow \boxed{\sin 18° = \frac{\sqrt{5}-1}{4}} \qquad (\because \ \sin 18 > 0)$$

B.

$$\cos 2\theta = 1 - 2\sin^2\theta$$

$$\Rightarrow \cos 36° = 1 - 2\sin^2 18$$

$$= 1 - 2\left(\frac{-1+\sqrt{5}}{4}\right)^2$$

$$= 1 - 2 \times \frac{1 - 2\sqrt{5} + 5}{16}$$

$$= 1 - \frac{3 - \sqrt{5}}{4}$$

$$\Rightarrow \boxed{\cos 36° = \frac{1+\sqrt{5}}{4}}$$

C.

$$\tan 2\theta = \frac{2\tan\theta}{1 - \tan^2\theta}$$

$$\Rightarrow \tan 30° = \frac{2\tan 15}{1 - \tan^2 15}$$

$$\Rightarrow \frac{1}{\sqrt{3}} = \frac{2\tan 15°}{1 - \tan^2 15°}$$

$$\Rightarrow 1 - \tan^2 15° = 2\sqrt{3} \cdot \tan 15°$$

$$\Rightarrow \tan^2 15° + 2\sqrt{3} \cdot \tan 15° - 1 = 0$$

$$\Rightarrow \tan 15° = \frac{-2\sqrt{3} \pm \sqrt{\left(2\sqrt{3}\right)^2 - 4 \times 1 \times (-1)}}{2 \times 1}$$

$$\Rightarrow \tan 15° = \frac{-2\sqrt{3} \pm \sqrt{12 + 4}}{2}$$

$$\Rightarrow \quad \tan 15° = -\sqrt{3} \pm 2$$

$$\Rightarrow \quad \boxed{\tan 15° = 2 - \sqrt{3}} \qquad (\because \ \tan 15 > 0)$$

Problem 2

Find the values of $\quad \tan^2\left(\dfrac{\pi}{16}\right) + \tan^2\left(\dfrac{3\pi}{16}\right) + \tan^2\left(\dfrac{5\pi}{16}\right) + \tan^2\left(\dfrac{7\pi}{16}\right)$

Solution

$$\tan^2\left(\frac{\pi}{16}\right) + \tan^2\left(\frac{3\pi}{16}\right) + \tan^2\left(\frac{5\pi}{16}\right) + \tan^2\left(\frac{7\pi}{16}\right)$$

$$= \tan^2\left(\frac{\pi}{16}\right) + \tan^2\left(\frac{3\pi}{16}\right) + \tan^2\left(\frac{\pi}{2} - \frac{3\pi}{16}\right) + \tan^2\left(\frac{\pi}{2} - \frac{\pi}{16}\right)$$

$$= \tan^2\left(\frac{\pi}{16}\right) + \tan^2\left(\frac{3\pi}{16}\right) + \cot^2\left(\frac{3\pi}{16}\right) + \cot^2\left(\frac{\pi}{16}\right)$$

$$= \tan^2\left(\frac{\pi}{16}\right) + \cot^2\left(\frac{\pi}{16}\right) + \tan^2\left(\frac{3\pi}{16}\right) + \cot^2\left(\frac{3\pi}{16}\right)$$

$$\tan^2\left(\frac{\pi}{16}\right) + \cot^2\left(\frac{\pi}{16}\right) = \left(\tan\left(\frac{\pi}{16}\right) + \cot\left(\frac{\pi}{16}\right)\right)^2 - 2\tan^2\left(\frac{\pi}{16}\right)\cot^2\left(\frac{\pi}{16}\right)$$

$$= \left(\frac{\sin\left(\frac{\pi}{16}\right)}{\cos\left(\frac{\pi}{16}\right)} + \frac{\cos\left(\frac{\pi}{16}\right)}{\sin\left(\frac{\pi}{16}\right)}\right)^2 - 2$$

$$= \left(\frac{\sin^2\left(\frac{\pi}{16}\right) + \cos^2\left(\frac{\pi}{16}\right)}{\cos\left(\frac{\pi}{16}\right)\sin\left(\frac{\pi}{16}\right)}\right)^2 - 2$$

$$= \left(\frac{1}{\cos\left(\frac{\pi}{16}\right)\sin\left(\frac{\pi}{16}\right)}\right)^2 - 2$$

$$\Rightarrow \tan^2\left(\frac{\pi}{16}\right) + \cot^2\left(\frac{\pi}{16}\right) = \frac{1}{\sin^2\left(\frac{\pi}{16}\right)\cos^2\left(\frac{\pi}{16}\right)} - 2$$

$$\Rightarrow \quad \tan^2\left(\frac{\pi}{16}\right) + \cot^2\left(\frac{\pi}{16}\right) = \frac{4}{\sin^2\left(\frac{\pi}{8}\right)} - 2 \qquad (\because \quad \sin 2x = 2\sin x \cos x)$$

Similarly

$$\tan^2\left(\frac{3\pi}{16}\right) + \cot^2\left(\frac{3\pi}{16}\right) = \frac{4}{\sin^2\left(\frac{3\pi}{8}\right)} - 2$$

$$\Rightarrow \quad \tan^2\left(\frac{3\pi}{16}\right) + \cot^2\left(\frac{3\pi}{16}\right) = \frac{4}{\sin^2\left(\frac{\pi}{2} - \frac{\pi}{8}\right)} - 2$$

$$\Rightarrow \quad \tan^2\left(\frac{3\pi}{16}\right) + \cot^2\left(\frac{3\pi}{16}\right) = \frac{4}{\cos^2\left(\frac{\pi}{8}\right)} - 2$$

So

$$\tan^2\left(\frac{\pi}{16}\right) + \tan^2\left(\frac{3\pi}{16}\right) + \tan^2\left(\frac{5\pi}{16}\right) + \tan^2\left(\frac{7\pi}{16}\right)$$

$$= \tan^2\left(\frac{\pi}{16}\right) + \cot^2\left(\frac{\pi}{16}\right) + \tan^2\left(\frac{3\pi}{16}\right) + \cot^2\left(\frac{3\pi}{16}\right)$$

$$= \frac{4}{\sin^2\left(\frac{\pi}{8}\right)} - 2 + \frac{4}{\cos^2\left(\frac{\pi}{8}\right)} - 2$$

$$= \frac{4\cos^2\left(\frac{\pi}{8}\right)}{\sin^2\left(\frac{\pi}{8}\right)} + \frac{4\sin^2\left(\frac{\pi}{8}\right)}{\cos^2\left(\frac{\pi}{8}\right)} - 4$$

$$= \frac{4}{\sin^2\left(\frac{\pi}{8}\right)\cos^2\left(\frac{\pi}{8}\right)} - 4$$

$$= \frac{4}{\sin^2\left(\frac{\pi}{4}\right)} - 4 \qquad (\because \quad \sin 2x = 2\sin x \cos x)$$

$$= \frac{16}{\frac{1}{2}} - 4 = 32 - 4$$

$$\Rightarrow \quad \boxed{\tan^2\left(\frac{\pi}{16}\right) + \tan^2\left(\frac{3\pi}{16}\right) + \tan^2\left(\frac{5\pi}{16}\right) + \tan^2\left(\frac{7\pi}{16}\right) = 28}$$

Problem 3

Find the values of $\left(1 - \cos\left(\frac{\pi}{8}\right)\right)\left(1 - \cos\left(\frac{3\pi}{8}\right)\right)\left(1 - \cos\left(\frac{5\pi}{8}\right)\right)\left(1 - \cos\left(\frac{7\pi}{8}\right)\right)$

Solution

$$\left(1 - \cos\left(\frac{\pi}{8}\right)\right)\left(1 - \cos\left(\frac{3\pi}{8}\right)\right)\left(1 - \cos\left(\frac{5\pi}{8}\right)\right)\left(1 - \cos\left(\frac{7\pi}{8}\right)\right)$$

$$= \left(1 - \cos\left(\frac{\pi}{8}\right)\right)\left(1 - \cos\left(\frac{7\pi}{8}\right)\right)\left(1 - \cos\left(\frac{3\pi}{8}\right)\right)\left(1 - \cos\left(\frac{5\pi}{8}\right)\right)$$

$$= \left(1 - \left[\cos\left(\frac{7\pi}{8}\right) + \cos\left(\frac{\pi}{8}\right)\right] + \cos\left(\frac{\pi}{8}\right)\cos\left(\frac{7\pi}{8}\right)\right)$$

$$\times \left(1 - \left[\cos\left(\frac{5\pi}{8}\right) + \cos\left(\frac{3\pi}{8}\right)\right] + \cos\left(\frac{5\pi}{8}\right)\cos\left(\frac{3\pi}{8}\right)\right)$$

We know $\cos(\pi - x) = -\cos x$ so

$$\cos\left(\frac{7\pi}{8}\right) + \cos\left(\frac{\pi}{8}\right) = -\cos\left(\frac{\pi}{8}\right) + \cos\left(\frac{\pi}{8}\right) = 0$$

$$\cos\left(\frac{5\pi}{8}\right) + \cos\left(\frac{3\pi}{8}\right) = -\cos\left(\frac{3\pi}{8}\right) + \cos\left(\frac{3\pi}{8}\right) = 0$$

$$\cos\left(\frac{\pi}{8}\right)\cos\left(\frac{7\pi}{8}\right) = -\cos^2\left(\frac{\pi}{8}\right)$$

$$\cos\left(\frac{5\pi}{8}\right)\cos\left(\frac{3\pi}{8}\right) = -\cos^2\left(\frac{3\pi}{8}\right)$$

Now

$$\left(1 - \cos\left(\frac{\pi}{8}\right)\right)\left(1 - \cos\left(\frac{3\pi}{8}\right)\right)\left(1 - \cos\left(\frac{5\pi}{8}\right)\right)\left(1 - \cos\left(\frac{7\pi}{8}\right)\right)$$

$$= \left(1 - 0 - \cos^2\left(\frac{\pi}{8}\right)\right) \times \left(1 - 0 - \cos^2\left(\frac{3\pi}{8}\right)\right)$$

$$= \sin^2\left(\frac{\pi}{8}\right)\sin^2\left(\frac{3\pi}{8}\right)$$

$$\sin x \sin y = -\frac{\cos(x+y) + \cos(x-y)}{2} \quad \text{then}$$

$$\left(1 - \cos\left(\frac{\pi}{8}\right)\right)\left(1 - \cos\left(\frac{3\pi}{8}\right)\right)\left(1 - \cos\left(\frac{5\pi}{8}\right)\right)\left(1 - \cos\left(\frac{7\pi}{8}\right)\right)$$

$$= \left(-\frac{\cos\left(\frac{\pi}{8} + \frac{3\pi}{8}\right) + \cos\left(\frac{\pi}{8} - \frac{3\pi}{8}\right)}{2}\right)^2$$

$$= \left(\frac{\cos\frac{\pi}{2} + \cos\left(-\frac{\pi}{4}\right)}{2}\right)^2$$

$$= \left(\frac{0 + \frac{1}{\sqrt{2}}}{2}\right)^2$$

$$\Rightarrow \boxed{\left(1 - \cos\left(\frac{\pi}{8}\right)\right)\left(1 - \cos\left(\frac{3\pi}{8}\right)\right)\left(1 - \cos\left(\frac{5\pi}{8}\right)\right)\left(1 - \cos\left(\frac{7\pi}{8}\right)\right) = \frac{1}{8}}$$

Problem 4

Find the value of $\quad \tan 9° - \tan 27° - \tan 63° + \tan 81°$

Solution

$\tan 9° - \tan 27° - \tan 63° + \tan 81°$

$$= (\tan 9° + \tan 81°) - (\tan 27° + \tan 63°)$$

$$= (\tan 9° + \cot 9°) - (\tan 27° + \cot 27°)$$

$$= \left(\frac{\sin 9°}{\cos 9°} + \frac{\cos 9°}{\sin 9°}\right) - \left(\frac{\sin 27°}{\cos 27°} + \frac{\cos 27°}{\sin 27°}\right)$$

$$= \left(\frac{\sin^2 9° + \cos^2 9°}{\cos 9° \cdot \sin 9°}\right) - \left(\frac{\sin^2 27° + \cos^2 27°}{\cos 27° \cdot \sin 27°}\right)$$

$$= \frac{1}{\cos 9° \cdot \sin 9°} - \frac{1}{\cos 27° \cdot \sin 27°}$$

$\tan 9° - \tan 27° - \tan 63° + \tan 81°$

$$= \frac{2}{2 \cdot \cos 9° \cdot \sin 9°} - \frac{2}{2 \cdot \cos 27° \cdot \sin 27°}$$

$$= \frac{2}{\sin 18°} - \frac{2}{\sin 54°}$$

$$= \frac{2}{\left(\frac{\sqrt{5}-1}{4}\right)} - \frac{2}{\left(\frac{\sqrt{5}+1}{4}\right)}$$

$$= \frac{8}{\sqrt{5}-1} - \frac{8}{\sqrt{5}+1}$$

$$= 8\left(\frac{1}{\sqrt{5}-1} - \frac{1}{\sqrt{5}+1}\right)$$

$$= 8\left(\frac{\sqrt{5}+1-\sqrt{5}+1}{5-1}\right)$$

$$= 8\left(\frac{2}{4}\right) = 4$$

$$\Rightarrow \boxed{\tan 9° - \tan 27° - \tan 63° + \tan 81° = 4}$$

Problem 5

Find the values of

A. $\sin\left(\dfrac{\cos^{-1}\left(\frac{1}{9}\right)}{2}\right)$

B. $\sin\left(\dfrac{\sin^{-1}\left(\frac{1}{9}\right)}{2}\right)$

C. $\sin\left(2\cos^{-1}\left(\frac{1}{9}\right)\right)$

Solution

A.

Let $\cos^{-1}\left(\frac{1}{9}\right) = 2x$ then $\cos 2x = \frac{1}{9}$

We know $\cos 2x = 1 - 2\cos^2 x$

so, $\frac{1}{9} = 1 - 2\cos^2 x$

$\Rightarrow \ 2\cos^2 x = 1 + \frac{1}{9} = \frac{10}{9}$

$\Rightarrow \ \cos^2 x = \frac{5}{9}$

$\sin^2 x + \cos^2 x = 1$ then

$\Rightarrow \ \sin^2 x = 1 - \cos^2 x = 1 - \frac{5}{9} = \frac{4}{9}$

$\Rightarrow \ \sin x = \frac{2}{3}$

$$\Rightarrow \quad \boxed{\sin\left(\frac{\cos^{-1}\left(\frac{1}{9}\right)}{2}\right) = \frac{2}{3}}$$

B.

Let $\sin^{-1}\left(\frac{1}{9}\right) = 2x$ then $\sin 2x = \frac{1}{9}$

$\sin^2 2x + \cos^2 2x = 1$

$\Rightarrow \ \frac{1}{81} + \cos^2 2x = 1$

$\Rightarrow \ \cos^2 2x = 1 - \frac{1}{81} = \frac{80}{81}$

$$\Rightarrow \quad \cos 2x = \frac{4\sqrt{5}}{9}$$

$$\cos 2x = 1 - 2\sin^2 x$$

$$\Rightarrow \quad \frac{4\sqrt{5}}{9} = 1 - 2\sin^2 x$$

$$\Rightarrow \quad \sin^2 x = \frac{1 - \frac{4\sqrt{5}}{9}}{2} = \frac{9 - 4\sqrt{5}}{18} = \frac{18 - 4\sqrt{20}}{18 \times 2} = \frac{10 + 8 - 4\sqrt{20}}{36}$$

$$\Rightarrow \quad \sin^2 x = \frac{10 - 4\sqrt{20} + 8}{36} = \frac{\left(\sqrt{10} - 2\sqrt{2}\right)^2}{36}$$

$$\Rightarrow \quad \sin x = \sqrt{\frac{\left(\sqrt{10} - 2\sqrt{2}\right)^2}{36}}$$

$$\Rightarrow \quad \boxed{\sin\left(\frac{\sin^{-1}\left(\frac{1}{9}\right)}{2}\right) = \frac{\sqrt{10} - 2\sqrt{2}}{6}}$$

C.

$$\sin\left(2\cos^{-1}\left(\frac{1}{9}\right)\right) = 2 \cdot \sin\left(\cos^{-1}\left(\frac{1}{9}\right)\right) \cdot \cos\left(\cos^{-1}\left(\frac{1}{9}\right)\right)$$

$$\Rightarrow \quad \sin\left(2\cos^{-1}\left(\frac{1}{9}\right)\right) = 2 \cdot \sqrt{1 - \cos^2\cos^{-1}\left(\frac{1}{9}\right)} \cdot \frac{1}{9}$$

$$= 2 \cdot \sqrt{1 - \left(\frac{1}{9}\right)^2} \cdot \frac{1}{9}$$

$$= 2 \cdot \frac{4\sqrt{5}}{9} \cdot \frac{1}{9}$$

$$\Rightarrow \quad \boxed{\sin\left(2\cos^{-1}\left(\frac{1}{9}\right)\right) = \frac{8\sqrt{5}}{81}}$$

Problem 6

$$\cos x + \sin x = \frac{1}{2} \quad \text{then} \quad \sin^6 x + \cos^6 x = ?$$

Solution

$$(\cos x + \sin x)^2 = \left(\frac{1}{2}\right)^2$$

$$\Rightarrow \cos^2 x + 2 \cdot \cos x \cdot \sin x + \sin^2 x = \frac{1}{4}$$

$$\Rightarrow \cos^2 x + \sin^2 x + 2 \cdot \cos x \cdot \sin x = \frac{1}{4}$$

$$\Rightarrow 1 + 2 \cdot \cos x \cdot \sin x = \frac{1}{4}$$

$$\Rightarrow 2 \cdot \cos x \cdot \sin x = \frac{1}{4} - 1$$

$$\Rightarrow \cos x \cdot \sin x = \frac{-3}{8}$$

$$\Rightarrow \sin^2 x \cdot \cos^2 x = \left(\frac{-3}{8}\right)^2 = \frac{9}{64}$$

$$(\sin^2 x + \cos^2 x)^3 = \sin^6 x + 3\sin^4 x \cdot \cos^2 x + 3\sin^2 x \cdot \cos^4 x + \cos^6 x$$

$$\Rightarrow 1 = \sin^6 x + 3\sin^4 x \cdot \cos^2 x + 3\sin^2 x \cdot \cos^4 x + \cos^6 x$$

$$\Rightarrow 1 = \sin^6 x + \cos^6 x + 3\sin^2 x \cdot \cos^2 x \cdot (\sin^2 x + \cos^2 x)$$

$$\Rightarrow 1 = \sin^6 x + \cos^6 x + 3\sin^2 x \cdot \cos^2 x \cdot 1$$

$$\Rightarrow \sin^6 x + \cos^6 x = 1 - 3\sin^2 x \cdot \cos^2 x$$

$$\Rightarrow \sin^6 x + \cos^6 x = 1 - 3 \times \frac{9}{64}$$

$$\Rightarrow \boxed{\sin^6 x + \cos^6 x = \frac{37}{64}}$$

Problem 7

Prove that:

A. $\sin^2 75° - \sin^2 15° = \dfrac{\sqrt{3}}{2}$

B. $\sin^2 48° - \cos^2 12° = -\dfrac{\sqrt{5}+1}{8}$

C. $(\sin 48° + \sin 18°)^2 + (\cos 48° + \cos 18°)^2 = 3$

Solution

A.

Methord 1

$\sin^2 75° - \sin^2 15° = (\sin 75° + \sin 15°)(\sin 75° - \sin 15°)$

We know

$$\sin x + \sin y = 2\sin\left(\frac{x+y}{2}\right)\cos\left(\frac{x-y}{2}\right)$$

$$\sin x - \sin y = 2\cos\left(\frac{x+y}{2}\right)\sin\left(\frac{x-y}{2}\right)$$

So

$\sin^2 75° - \sin^2 15°$

$$= 2\sin\left(\frac{75°+15°}{2}\right)\cos\left(\frac{75°-15°}{2}\right)\cdot 2\cos\left(\frac{75°+15°}{2}\right)\sin\left(\frac{75°-15°}{2}\right)$$

$\Rightarrow \sin^2 75° - \sin^2 15° = 2\sin 45° \cos 30° \cdot 2\cos 45° \sin 30°$

$$= 2\times\frac{1}{\sqrt{2}}\times\frac{\sqrt{3}}{2}\times 2\times\frac{1}{\sqrt{2}}\times\frac{1}{2}$$

$\Rightarrow \boxed{\sin^2 75° - \sin^2 15° = \dfrac{\sqrt{3}}{2}}$

Methord 2

We know

$$\sin(90 - x) = \cos x$$

So

$$\sin^2 75° - \sin^2 15° = \cos^2 15 - \sin^2 15°$$

$$\Rightarrow \quad \cos^2 15 - \sin^2 15° = \cos 30° \qquad\qquad (\because \cos 2x = \cos^2 x - \sin^2 x)$$

$$\Rightarrow \quad \sin^2 75° - \sin^2 15° = \cos 30°$$

$$\boxed{\Rightarrow \quad \sin^2 75° - \sin^2 15° = \frac{\sqrt{3}}{2}}$$

B.

$$\cos x = \sin(90 - x) \quad \text{so}$$

$$\sin^2 48° - \cos^2 12° = \sin^2 48° - \sin^2 78°$$

$$\Rightarrow \quad \sin^2 48° - \cos^2 12° = (\sin 48° + \sin 78°)(\sin 48° - \sin 78°)$$

We know

$$\sin x + \sin y = 2\sin\left(\frac{x+y}{2}\right)\cos\left(\frac{x-y}{2}\right)$$

$$\sin x - \sin y = 2\cos\left(\frac{x+y}{2}\right)\sin\left(\frac{x-y}{2}\right)$$

So

$$\sin^2 48° - \cos^2 12°$$

$$= 2\sin\left(\frac{48° + 78°}{2}\right)\cos\left(\frac{48° - 78°}{2}\right) \cdot 2\cos\left(\frac{48° + 78°}{2}\right)\sin\left(\frac{48° - 78°}{2}\right)$$

$$\Rightarrow \quad \sin^2 48° - \cos^2 12° = 2\sin 63° \cos(-15°) \cdot 2\cos 63° \sin(-15°)$$

$$= -2\sin 63° \cos 15° \cdot 2\cos 63° \sin 15°$$

$$= -2\sin 63° \cos 63° \cdot 2\sin 15° \cos 15°$$

$\Rightarrow \quad \sin^2 48° - \cos^2 12° = -\sin 126° \cdot \sin 30° \qquad (\because \ \sin 2x = 2\cos x \sin x)$

$$= -\cos 36° \cdot \sin 30°$$

$\Rightarrow \quad \sin^2 48° - \cos^2 12° = -\dfrac{\sqrt{5}+1}{4} \times \dfrac{1}{2}$

$\Rightarrow \quad \boxed{\sin^2 48° - \cos^2 12° = -\dfrac{\sqrt{5}+1}{8}}$

C.

$(\sin 48° + \sin 18°)^2 + (\cos 48° + \cos 18°)^2$

$$= (\sin^2 48° + 2 \cdot \sin 48° \cdot \sin 18° + \sin^2 18°)$$

$$+ (\cos^2 48° + 2 \cdot \cos 48° \cdot \cos 18° + \cos^2 18°)$$

$$= \sin^2 48° + \cos^2 48° + \sin^2 18° + \cos^2 18°$$

$$+ 2 \cdot \cos 48° \cdot \cos 6° + 2 \cdot \sin 48° \cdot \sin 18°$$

We know $\quad \sin^2 x + \cos^2 y = 1$

$2\cos x \cos y = \cos(x - y) + \cos(x + y)$

$2\sin x \sin y = \cos(x - y) - \cos(x + y)$

So

$(\sin 48° + \sin 18°)^2 + (\cos 48° + \cos 18°)^2$

$$= 1 + 1 + \cos(48° - 18°) + \cos(48° + 18°)$$

$$+ \cos(48° - 18°) - \cos(48° + 18°)$$

$$= 2 + \cos 30° + \cos 18° + \cos 30° - \cos 18°$$

$$= 2 + 2 \cdot \cos 30°$$

$$= 2 + 2 \cdot \dfrac{\sqrt{3}}{2}$$

$\Rightarrow \quad \boxed{(\sin 48° + \sin 18°)^2 + (\cos 48° + \cos 18°)^2 = 2 + \sqrt{3}}$

Problem 8

Find the value of
$$\frac{\sin^4\left(\frac{\pi}{8}\right) + \sin^4\left(\frac{3\pi}{8}\right) + \sin^4\left(\frac{5\pi}{8}\right) + \sin^4\left(\frac{7\pi}{8}\right)}{\cos^4\left(\frac{\pi}{8}\right) + \cos^4\left(\frac{3\pi}{8}\right) + \cos^4\left(\frac{5\pi}{8}\right) + \cos^4\left(\frac{7\pi}{8}\right)}$$

Solution

$$\sin^4\left(\frac{\pi}{8}\right) + \sin^4\left(\frac{3\pi}{8}\right) + \sin^4\left(\pi - \frac{3\pi}{8}\right) + \sin^4\left(\pi - \frac{\pi}{8}\right)$$

$$= \sin^4\left(\frac{\pi}{8}\right) + \sin^4\left(\frac{3\pi}{8}\right) + \sin^4\left(\frac{3\pi}{8}\right) + \sin^4\left(\frac{\pi}{8}\right)$$

$$= 2\sin^4\left(\frac{\pi}{8}\right) + 2\sin^4\left(\frac{3\pi}{8}\right)$$

$$= \sin^4\left(\frac{\pi}{8}\right) + \sin^4\left(\frac{\pi}{2} - \frac{\pi}{8}\right)$$

$$= 2\sin^4\left(\frac{\pi}{8}\right) + 2\cos^4\left(\frac{\pi}{8}\right)$$

$$\cos^4\left(\frac{\pi}{8}\right) + \cos^4\left(\frac{3\pi}{8}\right) + \cos^4\left(\frac{5\pi}{8}\right) + \cos^4\left(\frac{7\pi}{8}\right)$$

$$= \cos^4\left(\frac{\pi}{8}\right) + \cos^4\left(\frac{3\pi}{8}\right) + \cos^4\left(\frac{5\pi}{8}\right) + \cos^4\left(\frac{7\pi}{8}\right)$$

$$= \cos^4\left(\frac{\pi}{8}\right) + \cos^4\left(\frac{3\pi}{8}\right) + \cos^4\left(\pi - \frac{3\pi}{8}\right) + \cos^4\left(\pi - \frac{\pi}{8}\right)$$

$$= \cos^4\left(\frac{\pi}{8}\right) + \cos^4\left(\frac{3\pi}{8}\right) + \cos^4\left(\frac{3\pi}{8}\right) + \cos^4\left(\frac{\pi}{8}\right)$$

$$\Rightarrow \cos^4\left(\frac{\pi}{8}\right) + \cos^4\left(\frac{3\pi}{8}\right) + \cos^4\left(\frac{5\pi}{8}\right) + \cos^4\left(\frac{7\pi}{8}\right) = 2\cos^4\left(\frac{\pi}{8}\right) + 2\cos^4\left(\frac{3\pi}{8}\right)$$

$$= 2\cos^4\left(\frac{\pi}{8}\right) + 2\cos^4\left(\frac{\pi}{2} - \frac{\pi}{8}\right)$$

$$\Rightarrow \cos^4\left(\frac{\pi}{8}\right) + \cos^4\left(\frac{3\pi}{8}\right) + \cos^4\left(\frac{5\pi}{8}\right) + \cos^4\left(\frac{7\pi}{8}\right) = 2\cos^4\left(\frac{\pi}{8}\right) + 2\cos^4\left(\frac{\pi}{8}\right)$$

Now

$$\frac{\sin^4\left(\frac{\pi}{8}\right) + \sin^4\left(\frac{3\pi}{8}\right) + \sin^4\left(\frac{5\pi}{8}\right) + \sin^4\left(\frac{7\pi}{8}\right)}{\cos^4\left(\frac{\pi}{8}\right) + \cos^4\left(\frac{3\pi}{8}\right) + \cos^4\left(\frac{5\pi}{8}\right) + \cos^4\left(\frac{7\pi}{8}\right)} = \frac{2\cos^4\left(\frac{\pi}{8}\right) + 2\cos^4\left(\frac{\pi}{8}\right)}{2\cos^4\left(\frac{\pi}{8}\right) + 2\cos^4\left(\frac{\pi}{8}\right)}$$

$$\Rightarrow \boxed{\frac{\sin^4\left(\frac{\pi}{8}\right) + \sin^4\left(\frac{3\pi}{8}\right) + \sin^4\left(\frac{5\pi}{8}\right) + \sin^4\left(\frac{7\pi}{8}\right)}{\cos^4\left(\frac{\pi}{8}\right) + \cos^4\left(\frac{3\pi}{8}\right) + \cos^4\left(\frac{5\pi}{8}\right) + \cos^4\left(\frac{7\pi}{8}\right)} = 1}$$

Problem 9

Prove that: $1 - \cot 22° = \dfrac{2}{1 - \cot 23°}$

Solution

Let start with $(1 - \cot 22°)(1 - \cot 23°)$

$(1 - \cot 22°)(1 - \cot 23°)$

$$= 1 - \cot 23° - \cot 22° + \cot 22° \cdot \cot 23°$$

$$= 1 - \frac{\cos 23°}{\sin 23°} - \frac{\cos 22°}{\sin 22°} + \frac{\cos 23°}{\sin 23°} \cdot \frac{\cos 22°}{\sin 22°}$$

$$= \frac{\sin 22° \cdot \sin 23° - \cos 23° \cdot \sin 22° - \cos 22° \cdot \sin 23° + \cos 23° \cdot \cos 22°}{\sin 22° \cdot \sin 23°}$$

$$= \frac{\cos 23° \cdot \cos 22° + \sin 22° \cdot \sin 23° - (\cos 22° \cdot \sin 23° + \cos 23° \cdot \sin 22°)}{\sin 22° \cdot \sin 23°}$$

$$\Rightarrow (1 - \cot 22°)(1 - \cot 23°) = \frac{\cos 1° - \sin 45°}{\sin 22° \cdot \sin 23°}$$

$$= \frac{\sin 89° - \sin 45°}{\sin 22° \cdot \sin 23°}$$

$$\Rightarrow \quad (1 - \cot 22°)(1 - \cot 23°) = \frac{2\cos\left(\dfrac{89° + 45°}{2}\right)\sin\left(\dfrac{89° + 45°}{2}\right)}{\sin 22° \cdot \sin 23°}$$

$$= \frac{2\cos 67°\sin 22°}{\sin 22° \cdot \sin 23°}$$

$$= \frac{2\sin 23°}{\sin 23°}$$

$$\Rightarrow \quad (1 - \cot 22°)(1 - \cot 23°) = 2$$

$$\Rightarrow \quad \boxed{1 - \cot 22° = \frac{1}{1 - \cot 23°}}$$

Problem 10

If $\dfrac{\sin^4 x}{2} + \dfrac{\cos^4 x}{3} = \dfrac{1}{5}$ then find the value of $\sin^6 x + \cos^6 x$

Solution

$$\frac{\sin^4 x}{2} + \frac{\cos^4 x}{3} = \frac{1}{5}$$

$$\Rightarrow \quad \frac{5\sin^4 x}{2} + \frac{5\cos^4 x}{3} = 1$$

$$\Rightarrow \quad \left(1 + \frac{3}{2}\right)\sin^4 x + \left(1 + \frac{2}{3}\right)\cos^4 x = 1$$

$$\Rightarrow \quad \sin^4 x + \frac{3}{2} \times \sin^4 x + \cos^4 x + \frac{2}{3} \times \cos^4 x = 1$$

$$\Rightarrow \quad \sin^4 x + \cos^4 x + \frac{2}{3} \times \cos^4 x + \frac{3}{2} \times \sin^4 x = 1$$

$$(\sin^2 x + \cos^2 x)^2 = \sin^4 x + 2\sin^2 x \cdot \cos^2 x + \cos^4 x$$

$$\Rightarrow \quad 1 = \sin^4 x + \cos^4 x + 2\sin^2 x \cdot \cos^2 x$$

$$\Rightarrow \quad \sin^4 x + \cos^4 x = 1 - 2\sin^2 x \cdot \cos^2 x$$

So

$$\sin^4 x + \cos^4 x + \frac{2}{3} \times \cos^4 x + \frac{3}{2} \times \sin^4 x$$

$$= 1 - 2\sin^2 x \cdot \cos^2 x + \frac{2}{3} \times \cos^4 x + \frac{3}{2} \times \sin^4 x = 1$$

$$\Rightarrow \frac{2}{3} \times \cos^4 x + \frac{3}{2} \times \sin^4 x - 2\sin^2 x \cdot \cos^2 x = 0$$

$$\Rightarrow \left(\sqrt{\frac{2}{3}} \times \cos^2 x \right)^2 + \left(\sqrt{\frac{3}{2}} \times \sin^2 x \right)^2 - 2\sin^2 x \cdot \cos^2 x = 0$$

$$\Rightarrow \left(\sqrt{\frac{2}{3}} \times \cos^2 x - \sqrt{\frac{3}{2}} \times \sin^2 x \right)^2 = 0$$

$$\Rightarrow \sqrt{\frac{2}{3}} \times \cos^2 x - \sqrt{\frac{3}{2}} \times \sin^2 x = 0$$

$$\Rightarrow \sqrt{\frac{2}{3}} \times \cos^2 x = \sqrt{\frac{3}{2}} \times \sin^2 x$$

$$\Rightarrow 2\cos^2 x = 3\sin^2 x$$

$$\Rightarrow \frac{\sin^2 x}{\cos^2 x} = \frac{2}{3}$$

$$\Rightarrow \tan^2 x = \frac{2}{3}$$

$$\sec^2 x = 1 + \tan^2 x$$

$$\Rightarrow \sec^2 x = 1 + \frac{2}{3} = \frac{5}{3}$$

$$\Rightarrow \cos^2 x = \frac{3}{5}$$

$$\sin^2 x = 1 - \cos^2 x$$

$$\Rightarrow \quad \sin^2 x = 1 - \frac{3}{5} = \frac{2}{5}$$

$$\sin^6 x + \cos^6 x = (\sin^2 x)^3 + (\cos^2 x)^3$$

$$\Rightarrow \quad \sin^6 x + \cos^6 x = \left(\frac{2}{5}\right)^3 + \left(\frac{3}{5}\right)^3$$

$$\Rightarrow \quad \sin^6 x + \cos^6 x = \frac{8}{125} + \frac{27}{125} = \frac{35}{125}$$

$$\Rightarrow \quad \boxed{\sin^6 x + \cos^6 x = \frac{7}{25}}$$

Problem 11

If $A + B + C + D = 2\pi$ then Prove that

$$\frac{\tan A + \tan B + \tan C + \tan D}{\cot A + \cot B + \cot C + \cot D} = \tan A \cdot \tan B \cdot \tan C \cdot \tan D$$

Solution

$$A + B + C + D = 2\pi$$

$$\Rightarrow \quad A + B = 2\pi - (C + D)$$

$$\Rightarrow \quad \tan(A + B) = \tan\left(2\pi - (C + D)\right)$$

$$\Rightarrow \quad \tan(A + B) = -\tan(C + D)$$

$$\Rightarrow \quad \frac{\tan A + \tan B}{1 - \tan A \cdot \tan B} = -\frac{\tan C + \tan D}{1 - \tan C \cdot \tan D}$$

$$\Rightarrow \quad (\tan A + \tan B)(1 - \tan C \cdot \tan D) = -(\tan C + \tan D)(1 - \tan A \cdot \tan B)$$

$$\Rightarrow \quad \tan A - \tan A \cdot \tan C \cdot \tan D + \tan B - \tan B \cdot \tan C \cdot \tan D$$

$$= -\tan C + \tan A \cdot \tan B \cdot \tan C - \tan D + \tan A \cdot \tan B \cdot \tan D$$

$\Rightarrow \quad \tan A + \tan B + \tan C + \tan D$

$$= \tan A \cdot \tan C \cdot \tan D + \tan B \cdot \tan C \cdot \tan D$$

$$+ \tan A \cdot \tan B \cdot \tan C + \tan A \cdot \tan B \cdot \tan D$$

Divide with $\tan A \cdot \tan B \cdot \tan C \cdot \tan D$ then

$$\Rightarrow \frac{\tan A + \tan B + \tan C + \tan D}{\tan A \cdot \tan B \cdot \tan C \cdot \tan D}$$

$$= \frac{\tan A \cdot \tan C \cdot \tan D}{\tan A \cdot \tan B \cdot \tan C \cdot \tan D} + \frac{\tan B \cdot \tan C \cdot \tan D}{\tan A \cdot \tan B \cdot \tan C \cdot \tan D}$$

$$+ \frac{\tan A \cdot \tan B \cdot \tan C}{\tan A \cdot \tan B \cdot \tan C \cdot \tan D} + \frac{\tan A \cdot \tan B \cdot \tan D}{\tan A \cdot \tan B \cdot \tan C \cdot \tan D}$$

$$\Rightarrow \frac{\tan A + \tan B + \tan C + \tan D}{\tan A \cdot \tan B \cdot \tan C \cdot \tan D} = \frac{1}{\tan B} + \frac{1}{\tan A} + \frac{1}{\tan D} + \frac{1}{\tan C}$$

$$\Rightarrow \frac{\tan A + \tan B + \tan C + \tan D}{\tan A \cdot \tan B \cdot \tan C \cdot \tan D} = \cot A + \cot B + \cot C + \cot D$$

$$\Rightarrow \boxed{\begin{array}{c} \dfrac{\tan A + \tan B + \tan C + \tan D}{\cot A + \cot B + \cot C + \cot D} = \tan A \cdot \tan B \cdot \tan C \cdot \tan D \\[2ex] \textit{When} \quad A + B + C + D = 2\pi \end{array}}$$

Problem 12

If $A + B + C = \pi$, Then Prove that $\quad \sin\left(\dfrac{A}{2}\right) \cdot \sin\left(\dfrac{B}{2}\right) \cdot \sin\left(\dfrac{C}{2}\right) \geq \dfrac{1}{8}$

Solution

Let $x = \sin\left(\dfrac{A}{2}\right) \cdot \sin\left(\dfrac{B}{2}\right) \cdot \sin\left(\dfrac{C}{2}\right)$ then

$$x = \sin\left(\dfrac{A}{2}\right) \cdot \sin\left(\dfrac{B}{2}\right) \cdot \sin\left(\dfrac{C}{2}\right) = \dfrac{1}{2} \times \left(\cos\left(\dfrac{A-B}{2}\right) - \cos\left(\dfrac{A+B}{2}\right)\right) \cdot \sin\left(\dfrac{C}{2}\right)$$

We know $\quad A + B + C = \pi \quad \Rightarrow \quad A + B = \pi - C$

$$\Rightarrow \quad x = \dfrac{1}{2} \times \left(\cos\left(\dfrac{A-B}{2}\right) - \cos\left(\dfrac{\pi-C}{2}\right)\right) \cdot \sin\left(\dfrac{C}{2}\right)$$

$$\Rightarrow \quad x = \dfrac{1}{2} \times \left(\cos\left(\dfrac{A-B}{2}\right) - \sin\left(\dfrac{C}{2}\right)\right) \cdot \sin\left(\dfrac{C}{2}\right)$$

$$\Rightarrow \quad x = \dfrac{1}{2} \times \cos\left(\dfrac{A-B}{2}\right) \cdot \sin\left(\dfrac{C}{2}\right) - \dfrac{1}{2} \times \sin^2\left(\dfrac{C}{2}\right)$$

$$\Rightarrow \quad 2x = \cos\left(\dfrac{A-B}{2}\right) \cdot \sin\left(\dfrac{C}{2}\right) - \sin^2\left(\dfrac{C}{2}\right)$$

$$\Rightarrow \quad \sin^2\left(\dfrac{C}{2}\right) - \cos\left(\dfrac{A-B}{2}\right) \cdot \sin\left(\dfrac{C}{2}\right) + 2x = 0$$

Which is Quadratic Equation in $\sin C$

$$\Rightarrow \quad \sin\left(\dfrac{C}{2}\right) = \dfrac{\cos\left(\dfrac{A-B}{2}\right) \pm \sqrt{\left(\cos\left(\dfrac{A-B}{2}\right)\right)^2 - 4 \times 1 \times 2x}}{2 \times 1}$$

$$\Rightarrow \quad \sin\left(\dfrac{C}{2}\right) = \dfrac{\cos\left(\dfrac{A-B}{2}\right) \pm \sqrt{\cos^2\left(\dfrac{A-B}{2}\right) - 8x}}{2}$$

$\sin\left(\dfrac{C}{2}\right)$ has only real solutios, so $\quad \cos^2\left(\dfrac{A-B}{2}\right) - 8x \geq 0$

$$\Rightarrow \quad \cos^2\left(\frac{A-B}{2}\right) \geq 8x$$

$$\Rightarrow \quad 1 \geq 8x$$

$$\Rightarrow \quad x \leq \frac{1}{8}$$

$$\Rightarrow \quad \boxed{\begin{array}{c} \sin\left(\frac{A}{2}\right) \cdot \sin\left(\frac{B}{2}\right) \cdot \sin\left(\frac{C}{2}\right) \leq \frac{1}{8} \\[2mm] When \quad A + B + C = \pi \end{array}}$$

Problem 13

Prove that

A. If $A + B + C = \pi$ then $\tan A + \tan B + \tan C = \tan A \cdot \tan B \cdot \tan C$

B. If $A + B = \dfrac{\pi}{4}$ then $\tan A + \tan B + \tan A \cdot \tan B = 1$

C. If $A + B = \dfrac{\pi}{2}$ then $\tan A \cdot \tan B = 1$

Solution

A.

$$A + B + C = \pi$$

$$\Rightarrow \quad A + B = \pi - C$$

$$\Rightarrow \quad \tan(A + B) = \tan(\pi - C)$$

$$\Rightarrow \quad \frac{\tan A + \tan B}{1 - \tan A \cdot \tan B} = -\tan C$$

$$\Rightarrow \quad \tan A + \tan B = -\tan C + \tan A \cdot \tan B \cdot \tan C$$

$\Rightarrow \quad \tan A + \tan B + \tan C = \tan A \cdot \tan B \cdot \tan C$

B.

$A + B = \dfrac{\pi}{4}$

$\Rightarrow \quad \tan(A + B) = \tan\left(\dfrac{\pi}{4}\right)$

$\Rightarrow \quad \dfrac{\tan A + \tan B}{1 - \tan A \cdot \tan B} = 1$

$\Rightarrow \quad \tan A + \tan B + \tan A \cdot \tan B = 1$

C.

$A + B = \dfrac{\pi}{2}$

$\Rightarrow \quad A = \dfrac{\pi}{2} - B$

$\Rightarrow \quad \tan A = \tan\left(\dfrac{\pi}{2} - B\right)$

$\Rightarrow \quad \tan A = \cot B$

$\Rightarrow \quad \tan A \cdot \tan B = \cot B \cdot \tan B$

$\Rightarrow \quad \tan A \cdot \tan B = 1 \qquad (\because \cot B \cdot \tan B = 1)$

Problem 14

Find the minimum and maximum values of:

A. $f(x) = \sin^4 x + \cos^4 x$

B. $f(x) = \sin^6 x + \cos^6 x$

C. $f(x) = \sin^2 x + \cos^4 x$

Solution

A

$$\sin^4 x + \cos^4 x = (\sin^2 x)^2 + (\cos^2 x)^2$$

$$\Rightarrow f(x) = (\sin^2 x + \cos^2 x)^2 - 2\sin^2 x \cos^2 x$$

$$\Rightarrow f(x) = 1 - \frac{\sin^2 2x}{2} \qquad (\because \ \sin^2 x + \cos^2 x = 1 \ \& \ \sin x = 2\sin x \cos x)$$

$$\Rightarrow 0 \le \sin^2 2x \le 1$$

$$\Rightarrow 0 \le \frac{\sin^2 2x}{2} \le \frac{1}{2}$$

$$\Rightarrow 1 - 0 \ge 1 - \frac{\sin^2 2x}{2} \ge 1 - \frac{1}{2}$$

$$\Rightarrow 1 \ge f(x) \ge \frac{1}{2}$$

Hence

$$\boxed{\begin{array}{l} \text{maximum of } f(x) = 1 \\[2mm] \text{minimum of } f(x) = \dfrac{1}{2} \end{array}}$$

B

$$\sin^6 x + \cos^6 x = (\sin^2 x)^3 + (\cos^2 x)^3$$

$$\Rightarrow f(x) = (\sin^2 x + \cos^2 x)^3 - 3\sin^4 x \cos^2 x - 3\sin^2 x \cos^4 x$$

$$\Rightarrow f(x) = (\sin^2 x + \cos^2 x)^3 - 3\sin^2 x \cos^2 x (\sin^2 x + \cos^2 x)$$

$$\Rightarrow f(x) = 1 - 3 \times \frac{\sin^2 2x}{4} \qquad (\because \ \sin^2 x + \cos^2 x = 1 \ \& \ \sin x = 2\sin x \cos x)$$

$$\Rightarrow 0 \le \sin^2 2x \le 1$$

$$\Rightarrow 0 \le \frac{\sin^2 2x}{4} \le \frac{1}{4}$$

$$\Rightarrow \cdot 0 \le 3 \times \frac{\sin^2 2x}{4} \le \frac{3}{4}$$

$$\Rightarrow \quad 1 - 0 \geq 1 - 3 \times \frac{\sin^2 2x}{4} \geq 1 - \frac{3}{4}$$

$$\Rightarrow \quad 1 \geq f(x) \geq \frac{1}{4}$$

Hence

$$\boxed{\begin{array}{l} \text{maximum of } f(x) = 1 \\ \\ \text{minimum of } f(x) = \frac{1}{4} \end{array}}$$

C

$$\sin^2 x + \cos^4 x = \frac{1}{2}(2\sin^2 x) + \frac{1}{4}(4\cos^4 x)$$

$$\Rightarrow \quad f(x) = \frac{1}{2}(2\sin^2 x) + \frac{1}{4}(2\cos^2 x)^2$$

$$\Rightarrow \quad f(x) = \frac{1}{2}(1 - \cos 2x) + \frac{1}{4}(1 + \cos 2x)^2 \quad \{\cos x = 1 - 2\sin^2 x = 2\cos^2 x - 1\}$$

$$\Rightarrow \quad f(x) = \frac{1}{2}(1 - \cos 2x) + \frac{1}{4}(1 + 2\cos 2x + \cos^2 2x)$$

$$\Rightarrow \quad f(x) = \frac{1}{2} - \frac{\cos 2x}{2} + \frac{1}{4} + \frac{2\cos 2x}{4} + \frac{\cos^2 2x}{4}$$

$$\Rightarrow \quad f(x) = \frac{3}{4} + \frac{\cos^2 2x}{4}$$

$$\Rightarrow \quad 0 \leq \cos^2 2x \leq 1$$

$$\Rightarrow \quad 0 \leq \frac{\cos^2 2x}{4} \leq \frac{1}{4}$$

$$\Rightarrow \quad \frac{3}{4} + 0 \leq \frac{3}{4} + \frac{\cos^2 2x}{4} \leq \frac{3}{4} + \frac{1}{4}$$

$$\Rightarrow \quad \frac{3}{4} \leq f(x) \leq 1$$

Hence

$$\boxed{\begin{array}{l} \text{maximum of } f(x) = 1 \\ \\ \text{minimum of } f(x) = \frac{3}{4} \end{array}}$$

Problem 15

Simplify $\dfrac{\sin 2x + \sin 4x + \sin 6x + \sin 8x + \sin 10x}{\sin 7x + 2\sin 5x + \sin 3x}$, $x \neq \dfrac{n\pi}{2}, n \in Z$

Solution

$\dfrac{\sin 2x + \sin 4x + \sin 6x + \sin 8x + \sin 10x}{\sin 7x + 2\sin 5x + \sin 3x}$

$= \dfrac{(\sin 2x + \sin 8x) + (\sin 4x + \sin 6x) + \sin 10x}{(\sin 7x + \sin 3x) + 2\sin 5x}$

$\sin 2x + \sin 8x = 2\sin\left(\dfrac{2x + 8x}{2}\right)\cos\left(\dfrac{2x - 8x}{2}\right)$

$\Rightarrow \quad \sin 2x + \sin 8x = 2\sin 5x \cos 3x \qquad\qquad (\because \cos(-x) = \cos x)$

$\sin 4x + \sin 6x = 2\sin\left(\dfrac{4x + 6x}{2}\right)\cos\left(\dfrac{4x - 6x}{2}\right)$

$\Rightarrow \quad \sin 4x + \sin 6x = 2\sin 5x \cos x$

$\sin 10x = 2\sin 5x \cos 5x$

$\sin 7x + \sin 3x = 2\sin\left(\dfrac{7x + 3x}{2}\right)\cos\left(\dfrac{7x - 3x}{2}\right)$

$\Rightarrow \quad \sin 7x + \sin 3x = 2\sin 5x \cos 2x$

So

$\dfrac{\sin 2x + \sin 4x + \sin 6x + \sin 8x + \sin 10x}{\sin 7x + 2\sin 5x + \sin 3x}$

$= \dfrac{2\sin 5x \cos 3x + 2\sin 5x \cos x + 2\sin 5x \cos 5x}{2\sin 5x \cos 2x + 2\sin 5x}$

$= \dfrac{2\sin 5x}{2\sin 5x} \times \dfrac{\cos 3x + \cos x + \cos 5x}{\cos 2x + 1}$

$= \dfrac{(\cos 3x + \cos 5x) + \cos x}{\cos 2x + 1}$

$= \dfrac{2\cos\left(\dfrac{3x + 5x}{2}\right)\cos\left(\dfrac{3x - 5x}{2}\right) + \cos x}{\cos 2x + 1}$

$$\Rightarrow \frac{\sin 2x + \sin 4x + \sin 6x + \sin 8x + \sin 10x}{\sin 7x + 2\sin 5x + \sin 3x} = \frac{2\cos 4x \cos x + \cos x}{\cos 2x + 1}$$

$$= \frac{2\cos 4x \cos x + \cos x}{2\cos^2 x}$$

$$= \frac{2\cos 4x + 1}{2\cos x}$$

$$\Rightarrow \boxed{\frac{\sin 2x + \sin 4x + \sin 6x + \sin 8x + \sin 10x}{\sin 7x + 2\sin 5x + \sin 3x} = \frac{2\cos 4x + 1}{2\cos x}}$$

Trigonometric Equations

Problem 16

Solve for x: $\cos x \cdot \cos 2x \cdot \cos 3x = \dfrac{1}{4}$, $\quad x \in \left[-\dfrac{\pi}{2}, \dfrac{\pi}{2}\right]$

Solution

$\cos x \cdot \cos 2x \cdot \cos 3x = \dfrac{1}{4}$

$\Rightarrow 4 \cos 3x \cdot \cos x \cdot \cos 2x = 1$

$\Rightarrow 4 \left(\dfrac{\cos 2x + \cos 4x}{2}\right) \cdot \cos 2x = 1$

$\Rightarrow 2(\cos 2x + \cos 4x) \cdot \cos 2x = 1$

$\Rightarrow 2 \cos^2 2x + 2 \cos 4x \cdot \cos 2x = 1$

$\Rightarrow 2 \cos^2 2x + 2(2 \cos^2 2x - 1) \cdot \cos 2x = 1$

$\Rightarrow 2 \cos^2 2x + 4 \cos^3 2x - 2 \cos 2x = 1$

$\Rightarrow 2 \cos^2 2x + 4 \cos^3 2x - 2 \cos 2x - 1 = 0$

$\Rightarrow 2 \cos^2 2x\,(1 + 2 \cos 2x) - (1 + 2 \cos 2x) = 0$

$\Rightarrow (1 + 2 \cos 2x)(2 \cos^2 2x - 1) = 0$

$\Rightarrow (1 + 2 \cos 2x) \cdot \cos 4x = 0$

$\Rightarrow 1 + 2 \cos 2x = 0 \quad \text{or} \quad \cos 4x = 0$

If $\quad 1 + 2 \cos 2x = 0 \quad$ then

$\cos 2x = -\dfrac{1}{2}$

$\Rightarrow \cos 2x = \cos\left(\dfrac{2\pi}{3}\right)$

$\Rightarrow 2x = 2n\pi \pm \dfrac{2\pi}{3}$

$$\Rightarrow x = n\pi \pm \frac{\pi}{3}$$

$x \in \left[-\frac{\pi}{2}, \frac{\pi}{2}\right]$ then

$$x = \frac{\pi}{3}, \ -\frac{\pi}{3}$$

If $\cos 4x = 0$ then

$$\cos 4x = \cos\left(\frac{\pi}{2}\right)$$

$$\Rightarrow 4x = 2n\pi \pm \frac{\pi}{2}$$

$$\Rightarrow x = \frac{n\pi}{2} \pm \frac{\pi}{8}$$

$x \in \left[-\frac{\pi}{2}, \frac{\pi}{2}\right]$ then

$$x = \frac{\pi}{8}, \ -\frac{\pi}{8} \ \& \ -\frac{3\pi}{8}$$

$$\boxed{x = \frac{\pi}{3}, \ -\frac{\pi}{3}, \frac{\pi}{8}, \ -\frac{\pi}{8} \ -\frac{3\pi}{8}}$$

Problem 17

Find the value of x: $16^{\sin^2 x} + 16^{\cos^2 x} = 10$, $x \in \left[0, \frac{\pi}{2}\right]$

Solution

$$16^{\sin^2 x} + 16^{\cos^2 x} = 10$$

$\sin^2 x + \cos^2 x = 1$, so

$$16^{\sin^2 x} + 16^{\cos^2 x} = 16^{\sin^2 x} + 16^{1 - \sin^2 x}$$

$\Rightarrow \quad 16^{\sin^2 x} + \dfrac{16}{16^{\sin^2 x}} = 10$

Let $y = 16^{\sin^2 x}$ then

$$16^{\sin^2 x} + \dfrac{16}{16^{\sin^2 x}} = y + \dfrac{16}{y} = 10$$

$$\Rightarrow \dfrac{y^2 + 16}{y} = 10$$

$$\Rightarrow \ y^2 + 16 = 10y$$

$$\Rightarrow \ y^2 - 10y + 16 = 0$$

This is a quadratic equation with variable y.

So

$$y = \dfrac{-(-10) \pm \sqrt{(-10)^2 - 4 \times 1 \times 16}}{2 \times 1}$$

$$\Rightarrow \ y = \dfrac{10 \pm \sqrt{100 - 64}}{2} = \dfrac{10 \pm 6}{2} = 5 \pm 3$$

$$\Rightarrow \ y = 8, 2$$

When $y = 8$

$y = 16^{\sin^2 x} = 8$

$$\Rightarrow \ 2^{4 \cdot \sin^2 x} = 2^3$$

$$\Rightarrow \ 4 \cdot \sin^2 x = 3$$

$$\Rightarrow \ \sin^2 x = \dfrac{3}{4} = \left(\dfrac{\sqrt{3}}{2}\right)^2$$

$$\Rightarrow \ \sin^2 x = \sin^2\left(\dfrac{\pi}{3}\right)$$

$$\Rightarrow \ x = n\pi \pm \dfrac{\pi}{3}$$

$\Rightarrow \quad x = \dfrac{\pi}{3} \qquad \left(\because \ x \in \left[0, \dfrac{\pi}{2} \right] \right)$

When $y = 2$

$y = 16^{\sin^2 x} = 2$

$\Rightarrow \quad 2^{4 \cdot \sin^2 x} = 2^1$

$\Rightarrow \quad 4 \cdot \sin^2 x = 1$

$\Rightarrow \quad \sin^2 x = \dfrac{1}{4} = \left(\dfrac{1}{2} \right)^2$

$\Rightarrow \quad \sin^2 x = \sin^2 \left(\dfrac{\pi}{6} \right)$

$\Rightarrow \quad x = n\pi \pm \dfrac{\pi}{6}$

$\Rightarrow \quad x = \dfrac{\pi}{6} \qquad \left(\because \ x \in \left[0, \dfrac{\pi}{2} \right] \right)$

$\Rightarrow \quad \boxed{ x = \dfrac{\pi}{3}, \dfrac{\pi}{6} }$

Problem 18

Find the value of x : $3 \sin^4 x + \cos^4 x = 1$

Solution

$3 \sin^4 x + \cos^4 x = 1$

$1 = (\sin^2 x + \cos^2 x)^2$

$\Rightarrow \quad 1 = (\sin^2 x + \cos^2 x)^2$

$\Rightarrow \quad 1 = \sin^4 x + 2 \sin^2 x \cdot \cos^2 x + \cos^4 x$

$\Rightarrow \quad 3 \sin^4 x + \cos^4 x = \sin^4 x + 2 \sin^2 x \cdot \cos^2 x + \cos^4 x$

$\Rightarrow$ $2\sin^4 x - 2\sin^2 x \cdot \cos^2 x = 0$

$\Rightarrow$ $\sin^4 x - \sin^2 x \cdot \cos^2 x = 0$

$\Rightarrow$ $\sin^2 x \cdot (\sin^2 x - \cos^2 x) = 0$

$\Rightarrow$ $\sin^2 x \cdot (-\cos 2x) = 0$

$\Rightarrow$ $\sin x = 0$ or $\cos 2x = 0$

If $\sin x = 0$ then

$\sin x = \sin 0$

If $\sin x = \sin y$ then $x = n\pi + (-1)^n y$ so

$x = n\pi + (-1)^n 0 = n\pi$

If $\cos 2x = 0$ then

$\cos 2x = \cos\left(\dfrac{\pi}{2}\right)$

If $\cos x = \cos y$ then $x = 2n\pi \pm y$ so

$2x = 2n\pi \pm \dfrac{\pi}{2} = n\pi \pm \dfrac{\pi}{4}$

$\Rightarrow$ $\boxed{x = n\pi,\ n\pi \pm \dfrac{\pi}{4}}$

Problem 19

Solve for x: $\log_{|\cos x|} |\sin x| + \log_{|\sin x|} |\cos x| = 2$

Solution

$\log_{|\cos x|} |\sin x| + \log_{|\sin x|} |\cos x| = 2$

$\Rightarrow$ $\dfrac{\log |\sin x|}{\log |\cos x|} + \dfrac{\log |\cos x|}{\log |\sin x|} = 2$

$\Rightarrow (\log|\sin x|)^2 + (\log|\cos x|)^2 = 2\log|\sin x| \cdot \log|\cos x|$

$\Rightarrow (\log|\sin x|)^2 + (\log|\cos x|)^2 - 2\log|\sin x| \cdot \log|\cos x| = 0$

$\Rightarrow (\log|\sin x| - \log|\cos x|)^2 = 0$

$\Rightarrow \log|\sin x| - \log|\cos x| = 0$

$\Rightarrow \log|\sin x| = \log|\cos x|$

$\Rightarrow |\sin x| = |\cos x|$

$\Rightarrow \sin x = \pm \cos x$

When $\sin x = \cos x$

$\tan x = 1$

$\Rightarrow \tan x = \dfrac{\pi}{4}$

$\Rightarrow x = n\pi + \dfrac{\pi}{4}$ $\qquad (\because \tan x = \tan y \Rightarrow x = n\pi + y)$

When $\sin x = -\cos x$

$\tan x = -1$

$\Rightarrow \tan x = -\dfrac{\pi}{4}$

$\Rightarrow x = n\pi - \dfrac{\pi}{4}$ $\qquad (\because \tan x = \tan y \Rightarrow x = n\pi + y)$

In general $\boxed{x = n\pi \pm \dfrac{\pi}{4},\ n \in Z}$

Problem 20

Solve for x: $\cos 3x + \cos 2x = -1$

Solution

$\cos 3x + \cos 2x = -1$

$\Rightarrow 4\cos^3 x - 3\cos x + 2\cos^2 x - 1 = -1$

$\Rightarrow 4\cos^3 x - 3\cos x + 2\cos^2 x = 0$

$\Rightarrow \cos x\,(4\cos^2 x + 2\cos x - 3) = 0$

$\Rightarrow \cos x = 0 \quad \text{or} \quad 4\cos^2 x + 2\cos x - 3 = 0$

When $\cos x = 0$

$\cos x = \cos \dfrac{\pi}{2}$

$\Rightarrow x = \cos \dfrac{\pi}{2}$

$\Rightarrow x = 2n\pi \pm \dfrac{\pi}{2} \qquad\qquad (\because \cos x = \cos y \;\Rightarrow\; x = 2n\pi \pm y)$

When

$4\cos^2 x + 2\cos x - 3 = 0$

$\Rightarrow \cos x = \dfrac{-2 \pm \sqrt{4 + 4 \times 4 \times 3}}{2 \times 4}$

$\Rightarrow \cos x = \dfrac{-2 \pm 2\sqrt{13}}{2 \times 4} = \dfrac{-1 \pm \sqrt{13}}{4}$

$\Rightarrow \cos x = \dfrac{-1 + \sqrt{13}}{4} \qquad\qquad (\because -1 \leq \cos x \leq 1)$

$\Rightarrow \cos x = \cos^{-1}\left(\dfrac{-1 + \sqrt{13}}{4}\right)$

$\Rightarrow x = 2n\pi \pm \cos^{-1}\left(\dfrac{-1 + \sqrt{13}}{4}\right)$

$$x = 2n\pi \pm \frac{\pi}{2} \quad \text{or} \quad x = 2n\pi \pm \cos^{-1}\left(\frac{-1 + \sqrt{13}}{4}\right)$$

Problem 21

Solve for x: $4^{\cos x - 2\sin^2 x - 4\cos^3 x} = \frac{1}{8}$

Solution

$$4^{\cos x - 2\sin^2 x - 4\cos^3 x} = \frac{1}{8}$$

$$\Rightarrow (2^2)^{\cos x - 2\sin^2 x - 4\cos^3 x} = \frac{1}{2^3}$$

$$\Rightarrow 2^{2\cos x - 4\sin^2 x - 8\cos^3 x} = 2^{-3}$$

$$\Rightarrow 2\cos x - 4\sin^2 x - 8\cos^3 x = -3$$

$$\Rightarrow 2\cos x - 4(1 - \cos^2 x) - 8\cos^3 x = -3 \qquad (\because \ \sin^2 x + \cos^2 x = 1)$$

$$\Rightarrow 2\cos x - 4 + 4\cos^2 x - 8\cos^3 x = -3$$

$$\Rightarrow 2\cos x + 4\cos^2 x - 8\cos^3 x - 1 = 0$$

$$\Rightarrow 8\cos^3 x - 4\cos^2 x - 2\cos x + 1 = 0$$

$$\Rightarrow 4\cos^2 x (2\cos x - 1) - (2\cos x - 1) = 0$$

$$\Rightarrow (2\cos x - 1)(4\cos^2 x - 1) = 0$$

$$\Rightarrow 2\cos x - 1 = 0 \quad \text{or} \quad 4\cos^2 x - 1 = 0$$

When $2\cos x - 1 = 0$

$$\cos x = \frac{1}{2}$$

$\Rightarrow \cos x = \cos\left(\dfrac{\pi}{3}\right)$

$\Rightarrow x = 2n\pi \pm \dfrac{\pi}{3}$

When $4\cos^2 x - 1 = 0$

$\cos^2 x = \dfrac{1}{4}$

$\Rightarrow \cos x = \pm\dfrac{1}{2}$

$\Rightarrow \cos x = \pm\cos\left(\dfrac{\pi}{3}\right)$

$\Rightarrow \cos x = \cos\left(\dfrac{\pi}{3}\right), \ -\cos\left(\dfrac{\pi}{3}\right)$

If $\cos x = -\cos\left(\dfrac{\pi}{3}\right)$

$\Rightarrow \cos x = \cos\left(\pi - \dfrac{\pi}{3}\right)$

$\Rightarrow \cos x = \cos\left(\dfrac{2\pi}{3}\right)$

$\Rightarrow x = 2n\pi \pm \dfrac{2\pi}{3}$

If $\cos x = \cos\left(\dfrac{\pi}{3}\right)$

$x = 2n\pi \pm \dfrac{\pi}{3}$

$$\boxed{\quad x = 2n\pi \pm \dfrac{\pi}{3} \quad x = 2n\pi \pm \dfrac{2\pi}{3} \quad}$$

Problem 22

Solve for x: $\sec^6 x - \tan^6 x = 1$

Solution

$\sec^6 x - \tan^6 x = 1$

$\Rightarrow 1 + \tan^6 x = \sec^6 x$

$\Rightarrow (1 + \tan^2 x)^3 - 3\tan^2 x - 3\tan^4 x = \sec^6 x$

$\Rightarrow (1 + \tan^2 x)^3 - 3\tan^2 x(1 + \tan^2 x) = \sec^6 x$

$\Rightarrow \sec^6 x - 3\tan^2 x \sec^2 x = \sec^6 x$ $(\because \ 1 + \tan^2 x = \sec^2 x)$

$\Rightarrow \tan^2 x \sec^2 x = 0$

$\Rightarrow \tan^2 x = 0$ $(\because \ \sec x \in (-\infty, -1] \cup [1, \infty) \ \Rightarrow \ \sec x \neq 0)$

$\Rightarrow \tan x = \tan 0$

$\Rightarrow \boxed{x = n\pi}$

Problem 23

Find the principal solution of $\cot^2 x + \cos 2x = 1$

Solution

$\cos 2x = \cos^2 x - \sin^2 x$

$\Rightarrow \ \cos 2x = \dfrac{\cos^2 x - \sin^2 x}{\cos^2 x + \sin^2 x}$

Multiply by $\dfrac{\sin^2 x}{\sin^2 x}$ then

$\cos 2x \times \dfrac{\sin^2 x}{\sin^2 x} = \dfrac{\cos^2 x - \sin^2 x}{\cos^2 x + \sin^2 x} \times \dfrac{\sin^2 x}{\sin^2 x}$

$$\Rightarrow \quad \cos 2x = \frac{\dfrac{\cos^2 x - \sin^2 x}{\sin^2 x}}{\dfrac{\cos^2 x + \sin^2 x}{\sin^2 x}} = \frac{\dfrac{\cos^2 x}{\sin^2 x} - \dfrac{\sin^2 x}{\sin^2 x}}{\dfrac{\cos^2 x}{\sin^2 x} + \dfrac{\sin^2 x}{\sin^2 x}}$$

$$\Rightarrow \quad \cos 2x = \frac{\cot^2 x - 1}{\cot^2 x + 1}$$

Now

$$\cot^2 x + \cos 2x = \cot^2 x + \frac{\cot^2 x - 1}{\cot^2 x + 1}$$

$$\Rightarrow \quad \cot^2 x + \frac{\cot^2 x - 1}{\cot^2 x + 1} = 1$$

Let $y = \cot^2 x$ then

$$y + \frac{y - 1}{y + 1} = 1$$

$$\Rightarrow \quad y(y + 1) + y - 1 = y + 1$$

$$\Rightarrow \quad y^2 + y + y - 1 = y + 1$$

$$\Rightarrow \quad y^2 + y - 2 = 0$$

$$\Rightarrow \quad y = \frac{-1 \pm \sqrt{1 - 4 \times 1 \times (-2)}}{2 \times 1} = \frac{-1 \pm 3}{2}$$

$$\Rightarrow \quad y = 2$$

$$\Rightarrow \quad y = \cot^2 x = 2$$

$$\Rightarrow \quad y = \cot^2 x = 2$$

$$\Rightarrow \quad \cot x = \sqrt{2}$$

$$\Rightarrow \quad x = \cot^{-1} \sqrt{2}$$

$$\Rightarrow \quad \boxed{x = \frac{\pi}{4}}$$

Problem 24

Find the general solution of $\sin x + \sin 2x + \sin 3x = \cos x + \cos 2x + \cos 3x$

Solution

$\sin x + \sin 2x + \sin 3x = \cos x + \cos 2x + \cos 3x$

$\Rightarrow (\sin x + \sin 3x) + \sin 2x = (\cos x + \cos 3x) + \cos 2x$

$\Rightarrow 2 \sin \left(\dfrac{x + 3x}{2}\right) \cos \left(\dfrac{x - 3x}{2}\right) + \sin 2x = 2 \cos \left(\dfrac{x + 3x}{2}\right) \cos \left(\dfrac{x - 3x}{2}\right) + \cos 2x$

$\Rightarrow 2 \sin 2x \cos x + \sin 2x = 2 \cos 2x \cos x + \cos 2x$

$\Rightarrow \sin 2x (2 \cos x + 1) = \cos 2x (2 \cos x + 1)$

$\Rightarrow \sin 2x (2 \cos x + 1) - \cos 2x (2 \cos x + 1) = 0$

$\Rightarrow (2 \cos x + 1)(\sin 2x - \cos 2x) = 0$

$\Rightarrow 2 \cos x + 1 = 0 \quad$ or $\quad \sin 2x - \cos 2x = 0$

When $\quad 2 \cos x + 1 = 0$

$\cos x = -\dfrac{1}{2}$

$\Rightarrow \cos x = \cos \left(\dfrac{2\pi}{3}\right)$

$\Rightarrow x = 2n\pi \pm \dfrac{2\pi}{3}$

When $\quad \sin 2x - \cos 2x = 0$

$\sin 2x = \cos 2x$

$\Rightarrow \tan 2x = 1$

$\Rightarrow \tan 2x = \tan \left(\dfrac{\pi}{4}\right)$

$\Rightarrow 2x = n\pi \pm \dfrac{\pi}{4}$

$$\Rightarrow \quad x = \frac{n\pi}{2} \pm \frac{\pi}{8}$$

$$x = 2n\pi \pm \frac{2\pi}{3}$$

$$x = \frac{n\pi}{2} \pm \frac{\pi}{8}$$

$$\Rightarrow \quad x = \frac{\pi}{8}, \frac{2\pi}{3}, \frac{4\pi}{3}, \frac{13\pi}{8}, \dots\dots$$

Problem 25

Solve for x: $\sin 2x + \sin 6x + \sin 8x + \sin 10x + \sin 12x = 0$

Solution

$\sin 2x + \sin 4x + \sin 8x + \sin 10x + \sin 12x$

$\quad = (\sin 2x + \sin 10x) + (\sin 4x + \sin 8x) + \sin 12x$

$\quad = 2 \sin\left(\frac{2x + 10x}{2}\right) \cos\left(\frac{2x - 10x}{2}\right) + 2 \sin\left(\frac{4x + 8x}{2}\right) \cos\left(\frac{4x - 8x}{2}\right) + \sin 12x$

$\quad = 2 \sin 6x \cos 4x + 2 \sin 6x \cos 2x + 2 \sin 6x \cos 6x$

$\quad = 2 \sin 6x \left[\cos 4x + \cos 2x + \cos 6x\right]$

$\quad = 2 \sin 6x \left[\cos 4x + 2 \cos\left(\frac{2x + 6x}{2}\right) \cos\left(\frac{2x - 6x}{2}\right)\right]$

$\quad = 2 \sin 6x \left[\cos 4x + 2 \cos 4x \cos x\right]$

$\Rightarrow \quad 2 \sin 6x \cos 4x \left(1 + 2 \cos x\right) = 0$

$\Rightarrow \quad \sin 6x = 0, \quad \cos 4x = 0, \quad 1 + 2\cos x = 0$

When $\sin 6x = 0$

$\sin 6x = \sin 0$

$\Rightarrow \quad 6x = n\pi$

$$\Rightarrow \quad x = \frac{n\pi}{6}$$

When $\quad \cos 4x = 0$

$$\Rightarrow \quad 4x = 2n\pi \pm \frac{\pi}{2}$$

$$\Rightarrow \quad x = \frac{n\pi}{2} \pm \frac{\pi}{8}$$

When $\quad 1 + 2\cos x = 0$

$$\cos x = -\frac{1}{2}$$

$$\cos x = \cos\left(\frac{2\pi}{3}\right)$$

$$\Rightarrow \quad x = 2n\pi \pm \frac{2\pi}{3}$$

$$\Rightarrow \quad x = \frac{n\pi}{6} \pm \frac{2\pi}{3}$$

$$x = \frac{n\pi}{6}$$

$$x = \frac{n\pi}{2} \pm \frac{\pi}{8}$$

$$x = \frac{n\pi}{6} \pm \frac{2\pi}{3}$$

$$\Rightarrow \quad x = -\frac{2\pi}{3}, \; -\frac{\pi}{3}, \; -\frac{\pi}{6}, \; -\frac{\pi}{8}, \; 0, \; \frac{\pi}{8}, \frac{\pi}{6}, \frac{\pi}{3}, \ldots\ldots\ldots$$

Problem 26

Solve for x : $\dfrac{\sin x + \cos x}{\sin x - \cos x} = \sqrt{3}$

Solution

$\dfrac{\sin x + \cos x}{\sin x - \cos x} = \sqrt{3}$

Multiply and divide with $\dfrac{\cos x}{\cos x}$ then

$\Rightarrow \dfrac{\dfrac{\sin x + \cos x}{\cos x}}{\dfrac{\sin x - \cos x}{\cos x}} = \sqrt{3}$

$\Rightarrow \dfrac{\tan x + 1}{\tan x - 1} = \sqrt{3}$

$\Rightarrow \dfrac{1 + \tan x}{1 - \tan x} = -\sqrt{3}$

$\Rightarrow \dfrac{1 + \tan x}{1 - 1 \times \tan x} = \tan\left(-\dfrac{\pi}{3}\right)$

$\Rightarrow \dfrac{\tan\left(\dfrac{\pi}{4}\right) + \tan x}{1 - \tan\left(\dfrac{\pi}{4}\right) \times \tan x} = \tan\left(-\dfrac{\pi}{3}\right)$

$\Rightarrow \tan\left(\dfrac{\pi}{4} + x\right) = \tan\left(-\dfrac{\pi}{3}\right)$

$\Rightarrow \dfrac{\pi}{4} + x = n\pi - \dfrac{\pi}{3}$

$\Rightarrow \boxed{x = n\pi - \dfrac{7\pi}{12}}$

Problem 27

Find the smallest positive value of x :

$$\tan(x + 20°) = \tan(x - 10°) \cdot \tan x \cdot \tan(x + 10°)$$

Solution

$\tan(x + 20°) = \tan(x - 10°) \cdot \tan x \cdot \tan(x + 10°)$

$\Rightarrow \dfrac{\tan(x + 20°)}{\tan x} = \tan(x - 10°) \cdot \tan(x + 10°)$

$\Rightarrow \dfrac{\dfrac{\sin(x + 20°)}{\cos(x + 20°)}}{\dfrac{\sin x}{\cos x}} = \dfrac{\sin(x - 10°) \cdot \sin(x + 10°)}{\cos(x - 20°) \cdot \cos(x + 20°)}$

$\Rightarrow \dfrac{\sin(x + 20°) \cdot \cos x}{\cos(x + 20°) \cdot \sin x} = \dfrac{\sin(x - 10°) \cdot \sin(x + 10°)}{\cos(x - 10°) \cdot \cos(x + 10°)}$

If $\dfrac{a}{b} = \dfrac{c}{d}$ then $\dfrac{a + b}{a - b} = \dfrac{c + d}{c - d}$

So

$\dfrac{\sin(x + 20°) \cdot \cos x + \cos(x + 20°) \cdot \sin x}{\sin(x + 20°) \cdot \cos x - \cos(x + 20°) \cdot \sin x}$

$\qquad = \dfrac{\sin(x - 10°) \cdot \sin(x + 10°) + \cos(x - 10°) \cdot \cos(x + 10°)}{\sin(x - 10°) \cdot \sin(x + 10°) - \cos(x - 10°) \cdot \cos(x + 10°)}$

$\Rightarrow \dfrac{\sin(x + 20° + x)}{\sin(x + 20° - x)} = -\dfrac{\cos(x - 10° - x - 10°)}{\cos(x - 10° + x + 10°)}$

$\Rightarrow \dfrac{\sin(2x + 20°)}{\sin 20°} = -\dfrac{\cos(-20°)}{\cos 2x}$

$\Rightarrow \sin(2x + 20°) \cos 2x = -\cos 20° \sin 20°$

$\Rightarrow 2\sin(2x + 20°) \cos 2x = -2\cos 20° \sin 20°$

$\Rightarrow \sin(2x + 20° + 2x) + \sin(2x + 20° - 2x) = -\sin 40°$

$\Rightarrow \sin(4x + 20°) + \sin 20° = -\sin 40°$

$\Rightarrow \sin(4x + 20°) = -\sin 40° - \sin 20°$

$\Rightarrow \quad \sin(4x + 20°) = -(\sin 40° + \sin 20°)$

$\Rightarrow \quad \sin(4x + 20°) = -2\sin\left(\dfrac{40° + 20°}{2}\right)\cos\left(\dfrac{40° - 20°}{2}\right)$

$\Rightarrow \quad \sin(4x + 20°) = -2\sin 30°\cos 10°$

$\Rightarrow \quad \sin(4x + 20°) = -\cos 10°$

$\Rightarrow \quad \sin(4x + 20°) = \sin(-80°)$

$\Rightarrow \quad \sin(4x + 20°) = \sin(180 - (-80°))$ $\quad$ ($\because$ x is positive)

$\Rightarrow \quad \sin(4x + 20°) = \sin 260°$

$\Rightarrow \quad 4x + 20° = 260°$

$\Rightarrow \quad 4x = 240°$

$\Rightarrow \quad \boxed{x = 60°}$

Problem 28

Find the value of x: $(\cos x)^{2\sin^2 x - 3\sin x + 1} = 1$

Solution

We have 3 different cases

1) $\cos x = 1$

2) $2\sin^2 x - 3\sin x + 1 = 0$

3) $\cos x = -1$ $\;if\;$ $|\sin^2 x - 3\sin x + 1|$ is an even number

Case 1

$\cos x = 1$

If $\cos x = \cos y$ $\;then\;$ $x = 2n\pi \pm y$ $\;so$

$$\cos x = \cos 0$$

$$\Rightarrow \ x = 2n\pi \pm 0$$

$$\Rightarrow \ x = 2n\pi, \ n \in Z$$

Case 2

$$2\sin^2 x \ - \ 3\sin x \ + 1 = 0$$

If $\ y = \sin x$

$$2\sin^2 x \ - \ 3\sin x \ + 1 = 2y^2 - 3y + 1$$

$$\Rightarrow \ 2y^2 - 3y + 1 = 0$$

$$\Rightarrow \ y = \frac{3 \pm \sqrt{9 - 4 \times 2 \times 1}}{2 \times 2}$$

$$\Rightarrow \ y = \frac{3 \pm \sqrt{9 - 8}}{2 \times 2}$$

$$\Rightarrow \ y = \frac{3 \pm 1}{4}$$

$$\Rightarrow \ y = \frac{1}{2}, 1$$

$$\Rightarrow \ \sin x = \frac{1}{2}, 1$$

When $\ \sin x = \frac{1}{2}$

$$\sin x = \sin\left(\frac{\pi}{6}\right)$$

$$\Rightarrow \ x = n\pi + (-1)^n \cdot \frac{\pi}{6}, \ n \in Z$$

When $\ \sin x = 1$

$$\sin x = \sin\left(\frac{\pi}{2}\right)$$

$$\Rightarrow \ x = n\pi + (-1)^n \cdot \frac{\pi}{2}, \ n \in Z$$

Case 3

$\cos x = -1$

If $\cos x = \cos y$ then $x = 2n\pi \pm y$ so

$\cos x = \cos \pi$

$\Rightarrow x = 2n\pi \pm \pi$

$\sin^2 x - 3\sin x + 1 = \sin^2(2n\pi \pm \pi) - 3\sin(2n\pi \pm \pi) + 1$

$\Rightarrow \sin^2 x - 3\sin x + 1 = 0 - 0 + 1 = 1$

$\Rightarrow |\sin^2 x - 3\sin x + 1| \neq$ even

$\Rightarrow x \neq 2n\pi \pm \pi$

so

$$x = 2n\pi, \quad n \in Z$$

$$x = n\pi + (-1)^n \cdot \frac{\pi}{2}, \quad n \in Z$$

$$x = n\pi + (-1)^n \cdot \frac{\pi}{6}, \quad n \in Z$$

Problem 29

Prove that : $\dfrac{\tan 3x}{\tan x} = \tan\left(\dfrac{\pi}{3} + x\right)\tan\left(\dfrac{\pi}{3} - x\right)$

Solution

Let start with $\tan 3x$

$$\tan 3x = \frac{3\tan x - \tan^3 x}{1 - 3\tan^2 x}$$

$$= \frac{3\tan x - \tan^3 x}{1 - 3\tan^2 x}$$

$$\Rightarrow \quad \tan 3x = \tan x \cdot \frac{3 - \tan^2 x}{1 - 3\tan^2 x}$$

$$= \tan x \cdot \frac{\left(\sqrt{3} + \tan x\right)\left(\sqrt{3} - \tan x\right)}{\left(1 + \sqrt{3}\tan x\right)\left(1 - \sqrt{3}\tan x\right)} \quad \left(\because a^2 - b^2 = (a+b)(a-b)\right)$$

$$\Rightarrow \quad \frac{\tan 3x}{\tan x} = \frac{\left(\sqrt{3} + \tan x\right)\left(\sqrt{3} - \tan x\right)}{\left(1 + \sqrt{3}\tan x\right)\left(1 - \sqrt{3}\tan x\right)}$$

$$\Rightarrow \quad \frac{\tan 3x}{\tan x} = \frac{\sqrt{3} + \tan x}{1 - \sqrt{3}\tan x} \times \frac{\sqrt{3} - \tan x}{1 - \sqrt{3}\tan x}$$

$$\Rightarrow \quad \frac{\tan 3x}{\tan x} = \frac{\tan\left(\frac{\pi}{3}\right) + \tan x}{1 - \tan\left(\frac{\pi}{3}\right)\tan x} \times \frac{\tan\left(\frac{\pi}{3}\right) - \tan x}{1 - \tan\left(\frac{\pi}{3}\right)\tan x}$$

$$\tan(A + B) = \frac{\tan A + \tan B}{1 - \tan A \tan B} \quad \& \quad \tan(A - B) = \frac{\tan A - \tan B}{1 + \tan A \tan B}$$

Then

$$\frac{\tan\left(\frac{\pi}{3}\right) + \tan x}{1 - \tan\left(\frac{\pi}{3}\right)\tan x} \times \frac{\tan\left(\frac{\pi}{3}\right) - \tan x}{1 - \tan\left(\frac{\pi}{3}\right)\tan x} = \tan\left(\frac{\pi}{3} + x\right)\tan\left(\frac{\pi}{3} - x\right)$$

$$\Rightarrow \quad \frac{\tan 3x}{\tan x} = \tan\left(\frac{\pi}{3} + x\right)\tan\left(\frac{\pi}{3} - x\right)$$

Problem 30

Find the smallest positive value of x

$$\sin^2 x + \sin^4 x + \sin^6 x + \sin^8 x + \ldots\ldots \infty = 1$$

Solution

$\sin^2 x + \sin^4 x + \sin^6 x + \sin^8 x + \ldots\ldots \infty$ is an infinite series, so

$$\sin^2 x + \sin^4 x + \sin^6 x + \sin^8 x + \ldots\ldots \infty = \frac{\sin^2 x}{1 - \sin^2 x}$$

$$\Rightarrow \quad \sin^2 x + \sin^4 x + \sin^6 x + \sin^8 x + \ldots\ldots \infty = \frac{\sin^2 x}{\cos^2 x}$$

$$\Rightarrow \quad \sin^2 x + \sin^4 x + \sin^6 x + \sin^8 x + \ldots\ldots \infty = \tan^2 x$$

$$\Rightarrow \quad \tan^2 x = 1$$

$$\Rightarrow \quad \tan x = \pm 1$$

$$\Rightarrow \quad x = \pm \frac{\pi}{4}$$

Problem 31

Find the smallest positive value of x: $\quad 3^{\sin 2x \,+\, 2\cos^2 x} + 3^{1 - \sin 2x \,+\, 2\sin^2 x} = 28$

Solution

$$3^{\sin 2x \,+\, 2\cos^2 x} + 3^{1 - \sin 2x \,+\, 2\sin^2 x} = 28$$

$$\Rightarrow \quad 3^{\sin 2x \,+\, 2\cos^2 x} + 3^{1 \,-\, \sin 2x \,+\, 2 \,-\, 2\cos^2 x} = 28$$

$$\Rightarrow \quad 3^{\sin 2x \,+\, 2\cos^2 x} + 3^{3 \,-\, \sin 2x \,-\, 2\cos^2 x} = 28$$

$$\Rightarrow \quad 3^{\sin 2x \,+\, 2\cos^2 x} + \frac{3^3}{3^{\sin 2x \,+\, 2\cos^2 x}} = 28$$

Let $y = 3^{\sin 2x \,+\, 2\cos^2 x}$, then

$$3^{\sin 2x \,+\, 2\cos^2 x} + \frac{3^3}{3^{\sin 2x \,+\, 2\cos^2 x}} = y + \frac{27}{y} = 28$$

$$\Rightarrow \quad \frac{y^2 + 27}{y} = 28$$

$$\Rightarrow \quad y^2 + 27 = 28y$$

$$\Rightarrow \quad y^2 - 28y + 27 = 0$$

$$\Rightarrow \quad y = \frac{28 \pm \sqrt{(-28)^2 - 4 \times 1 \times 27}}{2 \times 1} = \frac{28 \pm \sqrt{784 - 108}}{2} = \frac{28 \pm \sqrt{676}}{2}$$

$$\Rightarrow \quad y = \frac{28 \pm 26}{2}$$

$$\Rightarrow \quad y = 27, 1$$

When $y = 27$, then

$$3^{\sin 2x + 2\cos^2 x} = 27 = 3^3$$

$$\Rightarrow \quad \sin 2x + 2\cos^2 x = 3$$

$$\Rightarrow \quad \sin 2x + 1 + \cos 2x = 3$$

$$\Rightarrow \quad \sin 2x + \cos 2x = 2 \quad \Rightarrow \quad \sin 2x = \cos 2x = 1 \qquad (\textit{It is not possible})$$

When $y = 1$, then

$$3^{\sin 2x + 2\cos^2 x} = 1 = 3^0$$

$$\Rightarrow \quad \sin 2x + 2\cos^2 x = 0$$

$$\Rightarrow \quad \sin 2x + 1 + \cos 2x = 0$$

$$\Rightarrow \quad \sin 2x + \cos 2x = -1$$

$$\Rightarrow \quad \frac{1}{\sqrt{2}} \times \sin 2x + \frac{1}{\sqrt{2}} \times \cos 2x = -\frac{1}{\sqrt{2}}$$

$$\Rightarrow \quad \cos\left(\frac{\pi}{4}\right) \times \sin 2x + \sin\left(\frac{\pi}{4}\right) \times \cos 2x = \sin\left(\frac{5\pi}{4}\right)$$

$$\Rightarrow \quad \sin\left(\frac{\pi}{4} + 2x\right) = \sin\left(\frac{5\pi}{4}\right)$$

$$\Rightarrow \quad \frac{\pi}{4} + 2x = \frac{5\pi}{4}$$

$$\Rightarrow \quad 2x = \pi$$

$$\Rightarrow \quad x = \frac{\pi}{2}$$

Problem 32

Solve for x : $\cos x = \cos 2x + \sin x$

Solution

$\cos x - \sin x = \cos 2x$

$\Rightarrow \quad \cos x - \sin x = \cos 2x$

$\Rightarrow \quad \cos x - \sin x = \cos^2 x - \sin^2 x$

$\Rightarrow \quad \cos x - \sin x = (\cos x - \sin x)(\cos x + \sin x)$

$\Rightarrow \quad \cos x - \sin x - (\cos x - \sin x)(\cos x + \sin x) = 0$

$\Rightarrow \quad (\cos x - \sin x)(1 - \cos x - \sin x) = 0$

$\Rightarrow \quad \cos x - \sin x = 0 \quad or \quad 1 - \cos x - \sin x = 0$

When $\quad \cos x - \sin x = 0$

$\cos x = \sin x$

$\Rightarrow \quad \dfrac{\sin x}{\cos x} = 1$

$\Rightarrow \quad \tan x = \tan\left(\dfrac{\pi}{4}\right)$

$\Rightarrow \quad x = n\pi + \dfrac{\pi}{4}, \quad n \in Z$

When $\quad 1 - \cos x - \sin x = 0$

$\cos x + \sin x = 1$

$\Rightarrow \quad \dfrac{1}{\sqrt{2}} \times \cos x + \dfrac{1}{\sqrt{2}} \times \sin x = \dfrac{1}{\sqrt{2}} \times 1$

$\Rightarrow \quad \cos\left(\dfrac{\pi}{4}\right) \times \cos x + \cos\left(\dfrac{\pi}{4}\right) \times \sin x = \cos\left(\dfrac{\pi}{4}\right)$

$\Rightarrow \quad \cos\left(\dfrac{\pi}{4} - x\right) = \cos\left(\dfrac{\pi}{4}\right)$

$$\Rightarrow \quad \frac{\pi}{4} - x = 2n\pi \pm \frac{\pi}{4}$$

When $\dfrac{\pi}{4} - x = 2n\pi + \dfrac{\pi}{4}$

$$x = -2n\pi$$

$$\Rightarrow \quad x = 2n\pi, \quad n \in Z$$

When $\dfrac{\pi}{4} - x = 2n\pi - \dfrac{\pi}{4}$

$$-x = 2n\pi - \frac{\pi}{2}$$

$$\Rightarrow \quad x = -2n\pi + \frac{\pi}{2}$$

$$\Rightarrow \quad x = 2n\pi + \frac{\pi}{2}, \quad n \in Z$$

So

$$\boxed{\begin{array}{l} x = n\pi + \dfrac{\pi}{4}, \quad n \in Z \\[2mm] x = 2n\pi, \quad n \in Z \\[2mm] x = 2n\pi + \dfrac{\pi}{2}, \quad n \in Z \end{array}}$$

Problem 33

Solve for x: $\tan x + \tan\left(x + \dfrac{\pi}{3}\right) + \tan\left(x + \dfrac{2\pi}{3}\right) = 3$

Solution

$$\tan x + \tan\left(x + \frac{\pi}{3}\right) + \tan\left(x + \frac{2\pi}{3}\right) = 3$$

$$\Rightarrow \quad \tan x + \frac{\tan x + \tan\left(\dfrac{\pi}{3}\right)}{1 - \tan x \cdot \tan\left(\dfrac{\pi}{3}\right)} + \frac{\tan x + \tan\left(\dfrac{2\pi}{3}\right)}{1 - \tan x \cdot \tan\left(\dfrac{\pi}{3}\right)} = 3$$

$\Rightarrow \quad \tan x + \dfrac{\tan x + \sqrt{3}}{1 - \tan x \cdot \sqrt{3}} + \dfrac{\tan x - \sqrt{3}}{1 + \tan x \cdot \sqrt{3}} = 3$

$\Rightarrow \quad \tan x + \dfrac{(\tan x + \sqrt{3})(1 + \tan x \cdot \sqrt{3}) + (\tan x - \sqrt{3})(1 - \tan x \cdot \sqrt{3})}{(1 - \tan x \cdot \sqrt{3})(1 + \tan x \cdot \sqrt{3})} = 3$

$\Rightarrow \quad \tan x + \dfrac{\tan x + \tan^2 x \cdot \sqrt{3} + \sqrt{3} + 3\tan x + \tan x - \tan^2 x \cdot \sqrt{3} - \sqrt{3} + 3\tan x}{1 - 3\tan^2 x} = 3$

$\Rightarrow \quad \tan x + \dfrac{8\tan x}{1 - 3\tan^2 x} = 3$

$\Rightarrow \quad \dfrac{\tan x\,(1 - 3\tan^2 x) + 8\tan x}{1 - 3\tan^2 x} = 3$

$\Rightarrow \quad \dfrac{\tan x - 3\tan^3 x + 8\tan x}{1 - 3\tan^2 x} = 3$

$\Rightarrow \quad \dfrac{9\tan x - 3\tan^3 x}{1 - 3\tan^2 x} = 3$

$\Rightarrow \quad \dfrac{3\tan x - \tan^3 x}{1 - 3\tan^2 x} = 1$

$\Rightarrow \quad \tan 3x = 1$

$\Rightarrow \quad \tan 3x = \tan\left(\dfrac{\pi}{4}\right)$

$\Rightarrow \quad 3x = n\pi + \dfrac{\pi}{4}$

$$\Rightarrow \quad \boxed{\; x = \dfrac{n\pi}{3} + \dfrac{\pi}{12}, \; n \in Z \;}$$

Inverse Trigonometric Functions

Problem 34

Solve for x: $\sin^{-1}\left(x + 1 + \dfrac{1}{x} + \dfrac{1}{x^2} + \ldots\right) + \cos^{-1}\left(1 + \dfrac{1}{x} + \dfrac{1}{x^2} + \dfrac{1}{x^3} + \ldots\right) = \dfrac{\pi}{2}$

Solution

If $\sin^{-1} A + \cos^{-1} B = \dfrac{\pi}{2}$ then $A = B$ so

$$\sin^{-1}\left(x + 1 + \frac{1}{x} + \frac{1}{x^2} + \ldots\right) + \cos^{-1}\left(1 + \frac{1}{x} + \frac{1}{x^2} + \frac{1}{x^3} + \ldots\right) = \frac{\pi}{2}$$

$$\Rightarrow \quad x + 1 + \frac{1}{x} + \frac{1}{x^2} + \ldots = 1 + \frac{1}{x} + \frac{1}{x^2} + \frac{1}{x^3} + \ldots$$

$x + 1 + \dfrac{1}{x} + \dfrac{1}{x^2} + \ldots$ and $1 + \dfrac{1}{x} + \dfrac{1}{x^2} + \dfrac{1}{x^3} + \ldots$ are infinite series so

$$x + 1 + \frac{1}{x} + \frac{1}{x^2} + \ldots = 1 + \frac{1}{x} + \frac{1}{x^2} + \frac{1}{x^3} + \ldots$$

$$\Rightarrow \quad \frac{x}{1 - \dfrac{1}{x}} = \frac{1}{1 - \dfrac{1}{x}}$$

$$\Rightarrow \quad \frac{x^2}{x - 1} = \frac{x}{x - 1}$$

$$\Rightarrow \quad \frac{x^2}{x - 1} - \frac{x}{x - 1} = 0$$

$$\Rightarrow \quad \frac{x}{x - 1}(x - 1) = 0$$

$$\Rightarrow \quad x = 0 \qquad\quad x - 1 \neq 0$$

When $x = 0$

$$\sin^{-1}\left(x + 1 + \frac{1}{x} + \frac{1}{x^2} + \ldots\right) = \sin^{-1}\left(0 + 1 + \frac{1}{0} + \frac{1}{0} + \ldots\right) = \sin^{-1}\infty$$

$\sin^{-1}\infty$ is not defind so x has no solution

Problem 35

Solve for x: $\sin^{-1} y + \cos^{-1} z = \dfrac{\pi}{2}$

$$y = \sqrt{x + \sqrt{x + \sqrt{x + ...}}}, \qquad y \geq 0$$

$$z = \sqrt{x^2 + \sqrt{x^2 + \sqrt{x^2 + ...}}} \qquad z \geq 0$$

Solution

Let   then

$$y = \sqrt{x + \sqrt{x + \sqrt{x + ...}}}$$

$$\Rightarrow y^2 = x + \sqrt{x + \sqrt{x + ...}}$$

$$\Rightarrow y^2 = x + y$$

$$\Rightarrow y^2 - y - x = 0$$

$$\Rightarrow y = \frac{1 \pm \sqrt{1 - 4 \times 1 \times (-x)}}{2 \times 1}$$

$$\Rightarrow y = \frac{1 \pm \sqrt{1 + 4x}}{2}$$

$$\Rightarrow y = \frac{1 + \sqrt{1 + 4x}}{2}, \; y \geq 0$$

$$z = \sqrt{x^2 + \sqrt{x^2 + \sqrt{x^2 + ...}}}$$

$$\Rightarrow z^2 = x^2 + \sqrt{x^2 + \sqrt{x^2 + ...}}$$

$$\Rightarrow z^2 = x^2 + z$$

$$\Rightarrow z^2 - z - x^2 = 0$$

$$\Rightarrow z = \frac{1 \pm \sqrt{1 - 4 \times 1 \times (-x^2)}}{2 \times 1}$$

$$\Rightarrow \quad z = \frac{1 \pm \sqrt{1 + 4x^2}}{2}$$

$$\Rightarrow \quad z = \frac{1 + \sqrt{1 + 4x^2}}{2}, \quad z \geq 0$$

$$\sin^{-1} y + \cos^{-1} z = \frac{\pi}{2}$$

$$\Rightarrow \quad y = z$$

$$\Rightarrow \quad \frac{1 + \sqrt{1 + 4x}}{2} = \frac{1 + \sqrt{1 + 4x^2}}{2}$$

$$\Rightarrow \quad \sqrt{1 + 4x} = \sqrt{1 + 4x^2}$$

$$\Rightarrow \quad 1 + 4x = 1 + 4x^2$$

$$\Rightarrow \quad 4x = 4x^2$$

$$\Rightarrow \quad 4x^2 - 4x = 0$$

$$\Rightarrow \quad 4x(x - 1) = 0$$

$$\Rightarrow \quad x = 0 \ \text{ or } \ x = 1$$

When $x = 1$

$$\sin^{-1} y + \cos^{-1} z \overset{?}{=} \frac{\pi}{2}$$

$$\Rightarrow \quad \sin^{-1}\left(\frac{1 + \sqrt{1 + 4}}{2}\right) + \cos^{-1}\left(\frac{1 + \sqrt{1 + 4}}{2}\right) \overset{?}{=} \frac{\pi}{2}$$

$$\Rightarrow \quad \sin^{-1}\left(\frac{1 + \sqrt{5}}{2}\right) + \cos^{-1}\left(\frac{1 + \sqrt{5}}{2}\right) \overset{?}{=} \frac{\pi}{2}$$

$\dfrac{1 + \sqrt{5}}{2} > 1 \ $ so $\ \sin^{-1}\left(\dfrac{1 + \sqrt{5}}{2}\right) \ $ & $\ \cos^{-1}\left(\dfrac{1 + \sqrt{5}}{2}\right) \ $ is not defind

$$\Rightarrow \quad \sin^{-1}\left(\frac{1 + \sqrt{5}}{2}\right) + \cos^{-1}\left(\frac{1 + \sqrt{5}}{2}\right) = \frac{\pi}{2}$$

$$\Rightarrow \quad x = 1 \ \text{is not a solution}$$

When $x = 0$

$$\sin^{-1} y + \cos^{-1} z \overset{?}{=} \frac{\pi}{2}$$

$$\Rightarrow \quad \sin^{-1}\left(\frac{1 + \sqrt{1 + 0}}{2}\right) + \cos^{-1}\left(\frac{1 + \sqrt{1 + 0}}{2}\right) \overset{?}{=} \frac{\pi}{2}$$

$$\Rightarrow \quad \sin^{-1}\left(\frac{2}{2}\right) + \cos^{-1}\left(\frac{2}{2}\right) \overset{?}{=} \frac{\pi}{2}$$

$$\Rightarrow \quad \sin^{-1} 1 + \cos^{-1} 1 \overset{?}{=} \frac{\pi}{2}$$

$$\Rightarrow \quad \frac{\pi}{2} + 0 = \frac{\pi}{2}$$

$$\Rightarrow \quad \sin^{-1} y + \cos^{-1} z = \frac{\pi}{2}$$

$$\Rightarrow \quad x = 0 \text{ is a solution}$$

Problem 36

Solve for x: $\sin^{-1} 4x + \sin^{-1} x = \dfrac{\pi}{4}$

Solution

$$\sin^{-1} 4x + \sin^{-1} x = \frac{\pi}{4}$$

$$\Rightarrow \quad \sin^{-1} 4x = \frac{\pi}{4} - \sin^{-1} x$$

$$\Rightarrow \quad \sin^{-1} 4x = \sin^{-1}\left(\frac{1}{\sqrt{2}}\right) - \sin^{-1} x$$

$$\Rightarrow \quad \sin^{-1} 4x = \sin^{-1}\left(\frac{1}{\sqrt{2}} \cdot \sqrt{1 - x^2} - x\sqrt{1 - \left(\frac{1}{\sqrt{2}}\right)^2}\right)$$

$$\Rightarrow \quad \sin^{-1} 4x = \sin^{-1}\left(\frac{1}{\sqrt{2}} \cdot \sqrt{1 - x^2} - x\sqrt{1 - \frac{1}{2}}\right)$$

$$\Rightarrow \quad \sin^{-1} 4x = \sin^{-1}\left(\frac{\sqrt{1-x^2}-x}{\sqrt{2}}\right)$$

$$\Rightarrow \quad 4x = \frac{\sqrt{1-x^2}-x}{\sqrt{2}}$$

$$\Rightarrow \quad 4\sqrt{2}x = \sqrt{1-x^2}-x$$

$$\Rightarrow \quad x\left(1+4\sqrt{2}\right) = \sqrt{1-x^2}$$

$$\Rightarrow \quad \left(x\left(1+4\sqrt{2}\right)\right)^2 = 1-x^2$$

$$\Rightarrow \quad x^2\left(1+8\sqrt{2}+32\right) = 1-x^2$$

$$\Rightarrow \quad x^2\left(8\sqrt{2}+33\right)+x^2 = 1$$

$$\Rightarrow \quad x^2 = \frac{1}{34+8\sqrt{2}}$$

$$\Rightarrow \quad x = \pm \frac{1}{\sqrt{34+8\sqrt{2}}}$$

When $x = \dfrac{-1}{\sqrt{34+8\sqrt{2}}}$ then sine values are negative, so

$$\sin^{-1}\left(\frac{-1}{\sqrt{34+8\sqrt{2}}}\right) + \sin^{-1}\left(\frac{-1}{\sqrt{34+8\sqrt{2}}}\right) \neq \frac{\pi}{4}$$

$$\Rightarrow \quad \boxed{x = \frac{1}{\sqrt{34+8\sqrt{2}}}}$$

Problem 37

Find the value of x : $\quad (\tan^{-1}x)^2 + (\cot^{-1}x)^2 = \dfrac{5\pi^2}{8}$

Solution

$$(\tan^{-1}x)^2 + (\cot^{-1}x)^2 = \frac{5\pi^2}{8}$$

$$\Rightarrow \ (\tan^{-1}x + \cot^{-1}x)^2 - 2\cdot \tan^{-1}x \cdot \cot^{-1}x = \frac{5\pi^2}{8}$$

$$\Rightarrow \ \left(\frac{\pi}{2}\right)^2 - 2\cdot \tan^{-1}x \cdot \left(\frac{\pi}{2} - \tan^{-1}x\right) = \frac{5\pi^2}{8}$$

$$\Rightarrow \ -\pi \cdot \tan^{-1}x + 2(\tan^{-1}x)^2 = \frac{3\pi^2}{8}$$

$$\Rightarrow \ 2(\tan^{-1}x)^2 - \pi \cdot \tan^{-1}x - \frac{3\pi^2}{8} = 0$$

$$\Rightarrow \ \tan^{-1}x = \frac{\pi \pm \sqrt{\pi^2 + 4 \times 2 \times \dfrac{3\pi^2}{8}}}{2 \times 2}$$

$$\Rightarrow \ \tan^{-1}x = \frac{\pi \pm \sqrt{\pi^2 + 3\pi^2}}{2 \times 2} = \frac{-\pi \pm 2\pi}{4}$$

$$\Rightarrow \ \tan^{-1}x \overset{?}{=} \frac{3\pi}{4}, \ -\frac{\pi}{4}$$

$$\Rightarrow \ \tan^{-1}x = -\frac{\pi}{4} \qquad\qquad \left(\because \ Range \ of \ \tan^{-1}x \ is \ \left(-\frac{\pi}{2}, \frac{\pi}{2}\right)\right)$$

$$\Rightarrow \ \cot^{-1}x = \frac{3\pi}{4} \qquad\qquad (\because \ Range \ of \ \cot^{-1}x \ is \ (0, \pi))$$

$$(\tan^{-1}x)^2 + (\cot^{-1}x)^2 \overset{?}{=} \frac{5\pi^2}{8}$$

$$\Rightarrow \ \left(-\frac{\pi}{4}\right)^2 + \left(\frac{3\pi}{4}\right)^2 \overset{?}{=} \frac{5\pi^2}{8}$$

$$\Rightarrow \ \frac{\pi^2}{16} + \frac{9\pi^2}{16} \overset{?}{=} \frac{5\pi^2}{8}$$

$$\Rightarrow \quad \frac{10\pi^2}{16} \stackrel{?}{=} \frac{5\pi^2}{8}$$

$$\Rightarrow \quad \frac{5\pi^2}{8} = \frac{5\pi^2}{8} \qquad \rightarrow \ x = -1 \ is \ a \ solution$$

$$\Rightarrow \quad \boxed{x = -1}$$

Problem 38

Solve for x: $\cos^{-1} x - \sin^{-1} x = \cos^{-1}\left(x\sqrt{3}\right)$

Solution

$$\cos^{-1} x - \sin^{-1} x = \cos^{-1}\left(x\sqrt{3}\right)$$

$$\Rightarrow \quad \cos^{-1} x - \left(\frac{\pi}{2} - \cos^{-1} x\right) = \cos^{-1}\left(x\sqrt{3}\right)$$

$$\Rightarrow \quad \cos^{-1} x - \frac{\pi}{2} + \cos^{-1} x = \cos^{-1}\left(x\sqrt{3}\right)$$

$$\Rightarrow \quad 2\cos^{-1} x = \frac{\pi}{2} + \cos^{-1}\left(x\sqrt{3}\right)$$

$$\Rightarrow \quad \cos^{-1}\left(2x^2 - 1\right) = \frac{\pi}{2} + \cos^{-1}\left(x\sqrt{3}\right)$$

$$\Rightarrow \quad \cos\left(\cos^{-1}\left(2x^2 - 1\right)\right) = \cos\left(\frac{\pi}{2} + \cos^{-1}\left(x\sqrt{3}\right)\right)$$

$$\Rightarrow \quad 2x^2 - 1 = \sin\cos^{-1}\left(x\sqrt{3}\right)$$

$$\Rightarrow \quad 2x^2 - 1 = -\sin\sin^{-1}\sqrt{1 - \left(x\sqrt{3}\right)^2}$$

$$\Rightarrow \quad 2x^2 - 1 = -\sqrt{1 - 3x^2}$$

$$\Rightarrow \quad \left(2x^2 - 1\right)^2 = 1 - 3x^2$$

$$\Rightarrow \quad 4x^4 - 4x^2 + 1 = 1 - 3x^2$$

$$\Rightarrow \quad 4x^4 - x^2 = 0$$

$$\Rightarrow \quad x^2\left(4x^2 - 1\right) = 0$$

$$\Rightarrow \ x^2(4x^2 - 1) = 0$$

$$\Rightarrow \ x^2(2x + 1)(2x - 1) = 0$$

$$\Rightarrow \ x^2 = 0, \ 2x + 1 = 0, \ 2x - 1 = 0$$

$$\Rightarrow \ x = 0, \ -\frac{1}{2}, \frac{1}{2}$$

Problem 39

Solve for x: $\ \sec^{-1}\left(\frac{x}{2}\right) - \sec^{-1}\left(\frac{x}{3}\right) = \sec^{-1} 3 - \sec^{-1} 2$

Solution

$$\sec^{-1}\left(\frac{x}{2}\right) - \sec^{-1}\left(\frac{x}{3}\right) = \sec^{-1} 3 - \sec^{-1} 2$$

$$\Rightarrow \ \cos^{-1}\left(\frac{2}{x}\right) - \cos^{-1}\left(\frac{3}{x}\right) = \cos^{-1}\left(\frac{1}{3}\right) - \cos^{-1}\left(\frac{1}{2}\right)$$

$$\Rightarrow \ \cos^{-1}\left(\frac{2}{x}\right) + \cos^{-1}\left(\frac{1}{2}\right) = \cos^{-1}\left(\frac{1}{3}\right) + \cos^{-1}\left(\frac{3}{x}\right)$$

$$\Rightarrow \cos^{-1}\left[\frac{2}{x} \cdot \frac{1}{2} - \sqrt{1 - \left(\frac{2}{x}\right)^2} \cdot \sqrt{1 - \left(\frac{1}{2}\right)^2}\right] = \cos^{-1}\left[\frac{1}{3} \cdot \frac{3}{x} - \sqrt{1 - \left(\frac{1}{3}\right)^2} \cdot \sqrt{1 - \left(\frac{3}{x}\right)^2}\right]$$

$$\Rightarrow \ \cos^{-1}\left[\frac{1}{x} - \sqrt{1 - \frac{4}{x^2}} \cdot \sqrt{1 - \frac{1}{4}}\right] = \cos^{-1}\left[\frac{1}{x} - \sqrt{1 - \frac{1}{9}} \cdot \sqrt{1 - \frac{9}{x^2}}\right]$$

$$\Rightarrow \ \frac{1}{x} - \sqrt{1 - \frac{4}{x^2}} \cdot \sqrt{1 - \frac{1}{4}} = \frac{1}{x} - \sqrt{1 - \frac{1}{9}} \cdot \sqrt{1 - \frac{9}{x^2}}$$

$$\Rightarrow \ \sqrt{1 - \frac{4}{x^2}} \cdot \sqrt{1 - \frac{1}{4}} = \sqrt{1 - \frac{1}{9}} \cdot \sqrt{1 - \frac{9}{x^2}}$$

$$\Rightarrow \ \sqrt{\frac{x^2 - 4}{x^2}} \cdot \sqrt{\frac{3}{4}} = \sqrt{\frac{8}{9}} \cdot \sqrt{\frac{x^2 - 9}{x^2}}$$

$$\Rightarrow \quad \frac{\sqrt{\dfrac{x^2-4}{x^2}}}{\sqrt{\dfrac{x^2-9}{x^2}}} = \frac{\sqrt{\dfrac{8}{9}}}{\sqrt{\dfrac{3}{4}}} = \sqrt{\frac{8}{9} \times \frac{4}{3}} = \sqrt{\frac{32}{27}}$$

$$\Rightarrow \quad \frac{x^2-4}{x^2-9} = \frac{32}{27}$$

$$\Rightarrow \quad 27x^2 - 108 = 32x^2 - 288$$

$$\Rightarrow \quad 5x^2 = 180$$

$$\Rightarrow \quad x^2 = 36$$

When $x = -6$

$$\sec^{-1}\left(\frac{x}{2}\right) - \sec^{-1}\left(\frac{x}{3}\right) = \sec^{-1}\left(\frac{-6}{2}\right) - \sec^{-1}\left(\frac{-6}{3}\right)$$

$$\Rightarrow \quad \sec^{-1}\left(\frac{x}{2}\right) - \sec^{-1}\left(\frac{x}{3}\right) = \sec^{-1}(-3) - \sec^{-1}(-2) \neq \sec^{-1} 3 - \sec^{-1} 2$$

$$\Rightarrow \quad x \neq -6$$

When $x = 6$

$$\sec^{-1}\left(\frac{x}{2}\right) - \sec^{-1}\left(\frac{x}{3}\right) = \sec^{-1}\left(\frac{6}{2}\right) - \sec^{-1}\left(\frac{6}{3}\right)$$

$$\Rightarrow \quad \sec^{-1}\left(\frac{x}{2}\right) - \sec^{-1}\left(\frac{x}{3}\right) = \sec^{-1}(3) - \sec^{-1}(2)$$

$$\Rightarrow \quad x = 6$$

so $\boxed{x = 6}$

Problem 40

Find the smallest positive value of $x : \sec^{-1}\left(\dfrac{2}{x}\right) + \sec^{-1} x = \sec^{-1} 2$

Solution

$\sec^{-1}\left(\dfrac{2}{x}\right) + \sec^{-1} x = \sec^{-1} 2$

$\Rightarrow \quad \sec^{-1}\left(\dfrac{2}{x}\right) = \sec^{-1} 2 - \sec^{-1} x$

$\Rightarrow \quad \cos^{-1}\left(\dfrac{x}{2}\right) = \cos^{-1}\left(\dfrac{1}{2}\right) - \cos^{-1}\left(\dfrac{1}{x}\right)$

$\Rightarrow \quad \cos^{-1}\left(\dfrac{x}{2}\right) = \cos^{-1}\left(\dfrac{1}{2}\times\dfrac{1}{x} + \sqrt{1-\left(\dfrac{1}{2}\right)^2}\cdot\sqrt{1-\left(\dfrac{1}{x}\right)^2}\right)$

$\Rightarrow \quad \cos^{-1}\left(\dfrac{x}{2}\right) = \cos^{-1}\left(\dfrac{1}{2x} + \sqrt{\dfrac{3}{4}}\cdot\sqrt{\dfrac{x^2-1}{x^2}}\right)$

$\Rightarrow \quad \dfrac{x}{2} = \dfrac{1}{2x} + \sqrt{\dfrac{3}{4}}\cdot\sqrt{\dfrac{x^2-1}{x^2}}$

$\Rightarrow \quad x = \dfrac{1}{x} + \sqrt{3}\sqrt{\dfrac{x^2-1}{x^2}}$

$\Rightarrow \quad x - \dfrac{1}{x} = \sqrt{3}\sqrt{\dfrac{x^2-1}{x^2}}$

$\Rightarrow \quad \left(x - \dfrac{1}{x}\right)^2 = 3 \times \dfrac{x^2-1}{x^2}$

$\Rightarrow \quad x^2 - 2 + \dfrac{1}{x^2} = 3 - \dfrac{3}{x^2}$

$\Rightarrow \quad x^2 + \dfrac{4}{x^2} - 5 = 0$

$\Rightarrow \quad \dfrac{x^4 + 4 - 5x^2}{x^2} = 0$

$\Rightarrow\ x^4 - 5x^2 + 4 = 0$

$\Rightarrow\ x^4 - x^2 - 4x^2 + 4 = 0$

$\Rightarrow\ x^2(x^2 - 1) - 4(x^2 - 1) = 0$

$\Rightarrow\ (x^2 - 1)(x^2 - 4) = 0$

$\Rightarrow\ x^2 - 1 = 0\ \ or\ \ x^2 - 4 = 0$

$\Rightarrow\ x = \pm 1\ \ or\ \ x = \pm 2$

When $x = 1$

$\sec^{-1}\left(\dfrac{2}{1}\right) + \sec^{-1} 1 \overset{?}{=} \sec^{-1} 2$

$\Rightarrow\ \sec^{-1} 2 + \sec^{-1} 1 \overset{?}{=} \sec^{-1} 2$

$\Rightarrow\ \dfrac{\pi}{3} + 0 \overset{?}{=} \dfrac{\pi}{3}$

$\Rightarrow\ \dfrac{\pi}{3} = \dfrac{\pi}{3}$

$\Rightarrow\ x = 1$ is a solution

When $x = -1$

$\sec^{-1}\left(\dfrac{2}{-1}\right) + \sec^{-1}(-1) \overset{?}{=} \sec^{-1} 2$

$\Rightarrow\ \sec^{-1}(-2) + \sec^{-1}(-1) \overset{?}{=} \sec^{-1} 2$

$\Rightarrow\ \dfrac{2\pi}{3} + \pi \overset{?}{=} \dfrac{\pi}{3}$

$\Rightarrow\ \dfrac{5\pi}{3} \neq \dfrac{\pi}{3}$

$\Rightarrow\ x = -1$ is not a solution

When $x = 2$

$\sec^{-1}\left(\dfrac{2}{2}\right) + \sec^{-1} 2 \overset{?}{=} \sec^{-1} 2$

$\Rightarrow \quad \sec^{-1} 1 + \sec^{-1} 2 \overset{?}{=} \sec^{-1} 2$

$\Rightarrow \quad 0 + \dfrac{\pi}{3} \overset{?}{=} \dfrac{\pi}{3}$

$\Rightarrow \quad \dfrac{\pi}{3} = \dfrac{\pi}{3}$

$\Rightarrow \quad x = 2$ is a solution

When $x = -2$

$\sec^{-1}\left(\dfrac{2}{-2}\right) + \sec^{-1}(-2) \overset{?}{=} \sec^{-1} 2$

$\Rightarrow \quad \sec^{-1}(-1) + \sec^{-1}(-2) \overset{?}{=} \sec^{-1} 2$

$\Rightarrow \quad \pi + \dfrac{2\pi}{3} \overset{?}{=} \dfrac{\pi}{3}$

$\Rightarrow \quad \dfrac{5\pi}{3} \neq \dfrac{\pi}{3}$

$\Rightarrow \quad x = -2$ is not a solution

So $\boxed{x = 1, 2}$

Properties of Triangle

Problem 41

In $\triangle ABC$, $\dfrac{\cos A}{a} = \dfrac{\cos B}{b} = \dfrac{\cos C}{c}$

then find the relation between

radius of the circumcircle and the

radius of the incircle

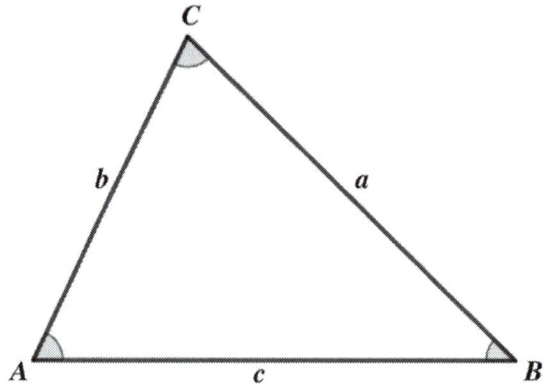

Solution

Let $R = radius\ of\ the\ circumcircle$ & $r = radius\ of\ the\ incircle$

$$\frac{\cos A}{a} = \frac{\cos B}{b} = \frac{\cos C}{c}$$

$$\frac{\cos A}{a} = \frac{\cos B}{b} \qquad \Rightarrow \qquad \frac{\cos A}{\cos B} = \frac{a}{b} \dots\dots\dots\dots\dots\dots\dots eq(1)$$

$$\frac{\cos B}{b} = \frac{\cos C}{c} \qquad \Rightarrow \qquad \frac{\cos B}{\cos C} = \frac{b}{c} \dots\dots\dots\dots\dots\dots\dots eq(2)$$

$$\frac{\sin A}{a} = \frac{\sin B}{b} = \frac{\sin C}{c} \quad (sine\ rule)$$

$$\frac{\sin A}{a} = \frac{\sin B}{b} \qquad \Rightarrow \qquad \frac{\sin A}{\sin B} = \frac{a}{b} \dots\dots\dots\dots\dots\dots\dots eq(3)$$

$$\frac{\sin B}{b} = \frac{\sin C}{c} \qquad \Rightarrow \qquad \frac{\sin B}{\sin C} = \frac{b}{c} \dots\dots\dots\dots\dots\dots\dots eq(4)$$

From $eq(1)$ & $eq(3)$

$$\frac{\cos A}{\cos B} = \frac{a}{b} = \frac{\sin A}{\sin B}$$

$\Rightarrow \quad \dfrac{\cos A}{\cos B} = \dfrac{\sin A}{\sin B}$

$\Rightarrow \quad \cos A \cdot \sin B = \sin A \cdot \cos B$

$\Rightarrow \quad \sin A \cdot \cos B - \cos A \cdot \sin B = 0$

$\Rightarrow \quad \sin(A - B) = 0$

$\Rightarrow \quad A - B = 0$

$\Rightarrow \quad A = B \dots\dots\dots\dots\dots\dots\dots\dots\dots\dots\dots\dots\dots\dots\dots eq(5)$

From $eq(3) \& eq(4)$

$\dfrac{\cos B}{\cos C} = \dfrac{b}{c} = \dfrac{\sin B}{\sin C}$

$\Rightarrow \quad \dfrac{\cos B}{\cos C} = \dfrac{\sin B}{\sin C}$

$\Rightarrow \quad \cos B \cdot \sin C = \sin B \cdot \cos C$

$\Rightarrow \quad \sin(B - C) = 0$

$\Rightarrow \quad B - C = 0$

$\Rightarrow \quad B = C \dots\dots\dots\dots\dots\dots\dots\dots\dots\dots\dots\dots\dots\dots\dots eq(6)$

From $eq(5) \& eq(6)$

$A = B = C \dots\dots\dots\dots\dots\dots\dots\dots\dots\dots\dots\dots\dots\dots\dots\dots eq(7)$

$A + B + C = 180°. \quad (ABC \text{ is a triangle}) \dots\dots\dots\dots\dots\dots\dots eq(8)$

From $eq(7) \& eq(8)$

$A = B = C = 60°$

$\Rightarrow \quad \Delta ABC$ is an equilateral triangle

$\Rightarrow \quad a = b = c$

$$r = \dfrac{2 \times Area\ of\ \Delta ABC}{Perimeter\ of\ \Delta ABC} = \dfrac{2 \times \frac{1}{2} \times a \times b \times \sin 60}{a + b + c} = \dfrac{a^2 \times \frac{\sqrt{3}}{2}}{3a}$$

$$\Rightarrow \quad r = \frac{a}{2\sqrt{3}}$$

$$2R = \frac{a}{\sin A} = \frac{a}{\sin 60} = \frac{a}{\frac{\sqrt{3}}{2}}$$

$$\Rightarrow \quad R = \frac{a}{\sqrt{3}}$$

$$\frac{r}{R} = \frac{\frac{a}{2\sqrt{3}}}{\frac{a}{\sqrt{3}}} = \frac{a}{2\sqrt{3}} \times \frac{\sqrt{3}}{a}$$

$$\Rightarrow \quad \boxed{\frac{r}{R} = \frac{1}{2}}$$

Problem 42

In ΔABC,

Prove that

$$\tan\left(\frac{B-C}{2}\right) = \frac{b-c}{b+c} \times \cot\left(\frac{A}{2}\right)$$

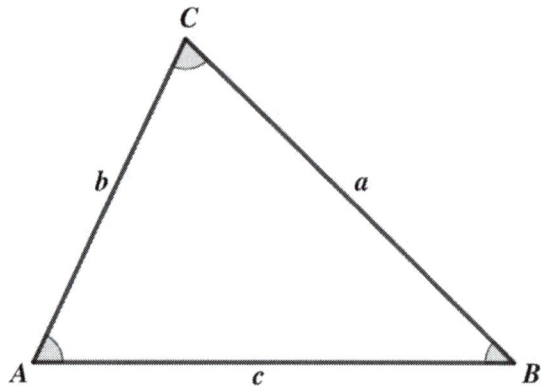

Solution

$$\tan\left(\frac{B-C}{2}\right) = \frac{\sin\left(\frac{B-C}{2}\right)}{\cos\left(\frac{B-C}{2}\right)}$$

$$\Rightarrow \quad \tan\left(\frac{B-C}{2}\right) = \frac{\sin\left(\frac{B-C}{2}\right)}{\cos\left(\frac{B-C}{2}\right)}$$

$$\Rightarrow \quad \tan\left(\frac{B-C}{2}\right) = \frac{\sin\left(\frac{B}{2}\right) \cdot \cos\left(\frac{C}{2}\right) - \cos\left(\frac{B}{2}\right) \cdot \sin\left(\frac{C}{2}\right)}{\cos\left(\frac{B}{2}\right) \cdot \cos\left(\frac{C}{2}\right) + \sin\left(\frac{B}{2}\right) \cdot \sin\left(\frac{C}{2}\right)}$$

$$\Rightarrow \quad \tan\left(\frac{B-C}{2}\right) = \frac{\sqrt{\frac{(s-a)(s-c)}{ac}} \cdot \sqrt{\frac{s(s-c)}{ab}} - \sqrt{\frac{s(s-b)}{ac}} \cdot \sqrt{\frac{(s-a)(s-b)}{ab}}}{\sqrt{\frac{s(s-b)}{ac}} \cdot \sqrt{\frac{s(s-c)}{ab}} + \sqrt{\frac{(s-a)(s-c)}{ac}} \cdot \sqrt{\frac{(s-a)(s-b)}{ab}}}$$

$$\Rightarrow \quad \tan\left(\frac{B-C}{2}\right) = \frac{\sqrt{(s-a)(s-c) \times s(s-c)} - \sqrt{s(s-b) \times (s-a)(s-b)}}{\sqrt{s(s-b) \times s(s-c)} + \sqrt{(s-a)(s-c) \times (s-a)(s-b)}}$$

$$\Rightarrow \quad \tan\left(\frac{B-C}{2}\right) = \frac{(s-c)\sqrt{s(s-a)} - (s-b)\sqrt{s(s-a)}}{s\sqrt{(s-b)(s-c)} + (s-a)\sqrt{(s-c)(s-b)}}$$

$$\Rightarrow \quad \tan\left(\frac{B-C}{2}\right) = \frac{(s-c-s+b)\sqrt{s(s-a)}}{(s+s-a)\sqrt{(s-b)(s-c)}}$$

$$\Rightarrow \quad \tan\left(\frac{B-C}{2}\right) = \frac{(b-c)\sqrt{s(s-a)}}{(2s-a)\sqrt{(s-b)(s-c)}}$$

$$\Rightarrow \quad \tan\left(\frac{B-C}{2}\right) = \frac{b-c}{a+b+c-a} \times \sqrt{\frac{s(s-a)}{(s-b)(s-c)}}$$

$$\Rightarrow \quad \tan\left(\frac{B-C}{2}\right) = \frac{b-c}{b+c} \times \cot\left(\frac{A}{2}\right)$$

Problem 43

In $\triangle ABC$,

$$\frac{1}{a+b} + \frac{1}{b+c} = \frac{1}{a+b} + \frac{1}{a+c} = \frac{3}{a+b+c}$$

then $\angle C = ?$

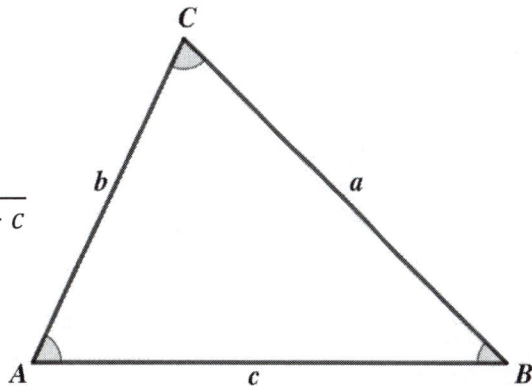

Solution

$$\frac{1}{a+b} + \frac{1}{b+c} = \frac{3}{a+b+c}$$

$$\Rightarrow \quad \frac{a+b+c}{a+b} + \frac{a+b+c}{b+c} = 3$$

$$\Rightarrow \quad 1 + \frac{c}{a+b} + \frac{a}{b+c} + 1 = 3$$

$$\Rightarrow \quad \frac{c}{a+b} + \frac{a}{b+c} = 1$$

$$\Rightarrow \quad \frac{c \cdot (b+c) + a \cdot (a+b)}{(a+b) \cdot (b+c)} = 1$$

$$\Rightarrow \quad c \cdot (b+c) + a \cdot (a+b) = (a+b) \cdot (b+c)$$

$$\Rightarrow \quad bc + c^2 + a^2 + ab = ab + ac + b^2 + bc$$

$$\Rightarrow \quad c^2 + a^2 = ac + b^2$$

$$\Rightarrow \quad c^2 + a^2 - b^2 = ac$$

$$\Rightarrow \quad b^2 = a^2 + c^2 - 2ac \cdot \cos B$$

$$\Rightarrow \quad \cos B = \frac{a^2 + c^2 - b^2}{2ac}$$

$$\Rightarrow \quad \cos B = \frac{a^2 + c^2 - b^2}{2ac} = \frac{a^2 + c^2 - b^2}{2(c^2 + a^2 - b^2)}$$

$$\Rightarrow \quad \cos B = \frac{1}{2}$$

$$\Rightarrow \quad \angle B = \frac{\pi}{3}$$

$$\frac{1}{a+b} + \frac{1}{b+c} = \frac{1}{a+b} + \frac{1}{a+c}$$

$$\Rightarrow \quad \frac{1}{b+c} = \frac{1}{a+c}$$

$$\Rightarrow \quad b+c = a+c$$

$$\Rightarrow \quad b = a$$

$$\Rightarrow \quad \angle B = \angle A = \frac{\pi}{3}$$

$$\Rightarrow \quad \angle C = \pi - (\angle B + \angle A)$$

$$\Rightarrow \quad \angle C = \pi - \left(\frac{\pi}{3} + \frac{\pi}{3}\right)$$

$$\Rightarrow \quad \boxed{\angle C = \frac{\pi}{3}}$$

Problem 44

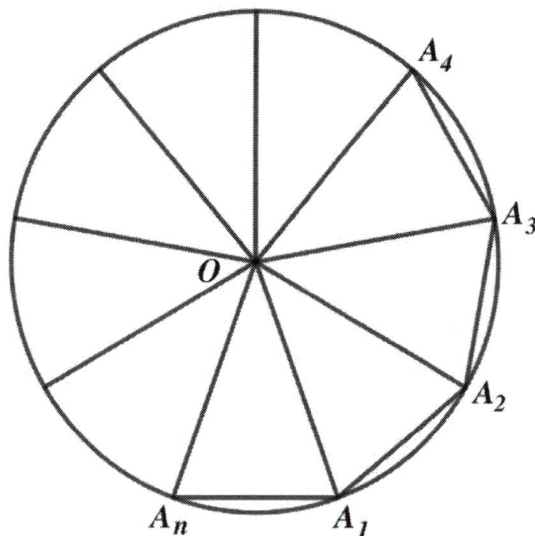

Let $A_1, A_2, A_3, \ldots A_n$ be the verticles of an $n -$ sided regular polygon

such that, $\dfrac{1}{A_1 A_2} = \dfrac{1}{A_1 A_3} + \dfrac{1}{A_1 A_4}$. Then find the value of n

Solution

$A_1, A_2, A_3, \ldots A_n$ is a regular polygon

So $OA_1 = OA_2 = OA_3 = \ldots = OA_n$ (Radius of the circumcircle)

Then, $\angle A_1 O A_2 = \angle A_2 O A_3 = \angle A_3 O A_4 = \ldots = \angle A_{n-1} O A_n = \angle A_n O A_1$

Let assume, $\angle A_1OA_2 = \angle A_2OA_3 = \angle A_3OA_4 = \dots = \angle A_{n-1}OA_n = \angle A_nOA_1 = x$

Then, $\angle A_1OA_2 = x$, $\angle A_1OA_3 = 2x$ & $\angle A_1OA_4 = 3x$

Polygon has n sides so, $x = \dfrac{2\pi}{n}$

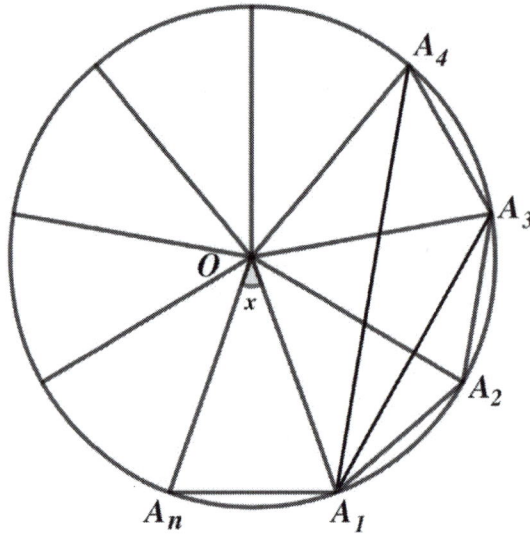

Let, Radius of the circumcircle $= r$, then

$OA_1 = OA_2 = OA_3 = \dots = OA_n = r$

From ΔOA_1A_2

$A_1A_2{}^2 = OA_1{}^2 + OA_2{}^2 - 2 \cdot OA_1 \cdot OA_2 \cdot \cos \angle A_1OA_2$

$\Rightarrow A_1A_2{}^2 = r^2 + r^2 - 2 \cdot r \cdot r \cdot \cos x$

$\Rightarrow A_1A_2{}^2 = 2r^2 - 2r^2 \cdot \cos x$

$\Rightarrow A_1A_2{}^2 = 2r^2(1 - \cos x)$

$\Rightarrow A_1A_2{}^2 = 2r^2 \times 2\sin^2\left(\dfrac{x}{2}\right)$

$\Rightarrow A_1A_2 = 2r \cdot \sin\left(\dfrac{x}{2}\right)$

$A_1A_3{}^2 = OA_1{}^2 + OA_3{}^2 - 2 \cdot OA_1 \cdot OA_3 \cdot \cos \angle A_1OA_3$

$\Rightarrow A_1A_3{}^2 = r^2 + r^2 - 2 \cdot r \cdot r \cdot \cos 2x$

$\Rightarrow A_1A_3{}^2 = 2r^2 - 2r^2 \cdot \cos 2x$

$\Rightarrow A_1A_3{}^2 = 2r^2(1 - \cos 2x)$

$\Rightarrow\ A_1A_3{}^2 = 2r^2 \times 2\sin^2 x$

$\Rightarrow\ A_1A_3 = 2r \cdot \sin x$

$A_1A_4{}^2 = OA_1{}^2 + OA_4{}^2 - 2 \cdot OA_1 \cdot OA_4 \cdot \cos \angle A_1OA_4$

$\Rightarrow\ A_1A_4{}^2 = r^2 + r^2 - 2 \cdot r \cdot r \cdot \cos 3x$

$\Rightarrow\ A_1A_4{}^2 = 2r^2 - 2r^2 \cdot \cos 3x$

$\Rightarrow\ A_1A_4{}^2 = 2r^2(1 - \cos 3x)$

$\Rightarrow\ A_1A_4{}^2 = 2r^2 \times 2\sin^2\left(\dfrac{3x}{2}\right)$

$\Rightarrow\ A_1A_4 = 2r \cdot \sin\left(\dfrac{3x}{2}\right)$

We know $\dfrac{1}{A_1A_2} = \dfrac{1}{A_1A_3} + \dfrac{1}{A_1A_4}$ then

$\dfrac{1}{2r \cdot \sin\left(\dfrac{x}{2}\right)} = \dfrac{1}{2r \cdot \sin x} + \dfrac{1}{2r \cdot \sin\left(\dfrac{3x}{2}\right)}$

$\Rightarrow\ \dfrac{1}{\sin\left(\dfrac{x}{2}\right)} = \dfrac{1}{\sin x} + \dfrac{1}{\sin\left(\dfrac{3x}{2}\right)}$

$\Rightarrow\ \dfrac{1}{\sin x} = \dfrac{1}{\sin\left(\dfrac{x}{2}\right)} - \dfrac{1}{\sin\left(\dfrac{3x}{2}\right)}$

$\Rightarrow\ \dfrac{1}{\sin x} = \dfrac{\sin\left(\dfrac{3x}{2}\right) - \sin\left(\dfrac{x}{2}\right)}{\sin\left(\dfrac{3x}{2}\right) \cdot \sin\left(\dfrac{x}{2}\right)}$

$\Rightarrow\ \dfrac{1}{\sin x} = \dfrac{2 \cdot \cos\left(\dfrac{\frac{3x}{2} + \frac{x}{2}}{2}\right) \cdot \sin\left(\dfrac{\frac{3x}{2} - \frac{x}{2}}{2}\right)}{\sin\left(\dfrac{3x}{2}\right) \cdot \sin\left(\dfrac{x}{2}\right)}$

$$\Rightarrow \quad \frac{1}{\sin x} = \frac{2 \cdot \cos x \cdot \sin\left(\frac{x}{2}\right)}{\sin\left(\frac{3x}{2}\right) \cdot \sin\left(\frac{x}{2}\right)}$$

$$\Rightarrow \quad \frac{1}{\sin x} = \frac{2 \cdot \cos x}{\sin\left(\frac{3x}{2}\right)}$$

$$\Rightarrow \quad \sin\left(\frac{3x}{2}\right) = 2 \cdot \cos x \cdot \sin x$$

$$\Rightarrow \quad \sin\left(\frac{3x}{2}\right) = \sin 2x$$

$$\Rightarrow \quad \sin\left(\frac{3x}{2}\right) = \sin(\pi - 2x)$$

$$\Rightarrow \quad \frac{3x}{2} = \pi - 2x$$

$$\Rightarrow \quad \frac{7x}{2} = \pi$$

$$\Rightarrow \quad x = \frac{2\pi}{7}$$

$$\Rightarrow \quad \frac{2\pi}{7} = \frac{2\pi}{n}$$

$$\Rightarrow \quad \boxed{\; n = 7 \;}$$

Problem 45

In $\triangle ABC$

$PA \perp BC, RC \perp AB$ & $BQ \perp AC$ and

$AP : CR : BQ = 3 : 4 : 5$

Find the value of $\sin A : \sin B : \sin C$

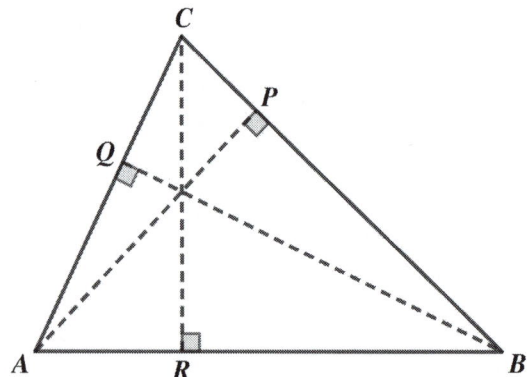

Solution

From $\triangle ABC$

$$\sin A = \frac{CR}{AC}$$

$$\Rightarrow AC = \frac{CR}{\sin A}$$

$$\sin B = \frac{AP}{AB}$$

$$\Rightarrow AB = \frac{AP}{\sin B}$$

$$\sin C = \frac{BQ}{BC}$$

$$\Rightarrow BC = \frac{BQ}{\sin C}$$

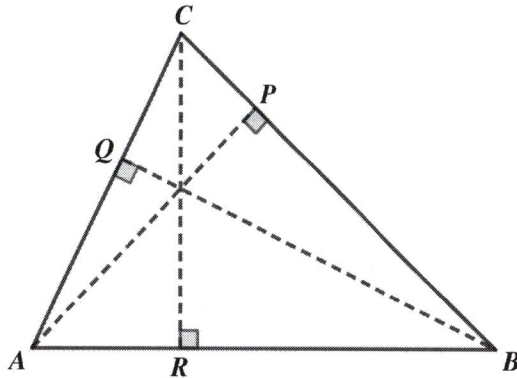

$$\frac{\sin A}{BC} = \frac{\sin B}{AC} = \frac{\sin C}{AB} \qquad (sine\ rule)$$

$$\frac{\sin A}{BC} = \frac{\sin B}{AC}$$

$$\Rightarrow \frac{\sin A}{\sin B} = \frac{BC}{AC}$$

$$\Rightarrow \frac{\sin A}{\sin B} = \frac{\dfrac{BQ}{\sin C}}{\dfrac{CR}{\sin A}}$$

$$\Rightarrow \frac{\sin A}{\sin B} = \frac{BQ \cdot \sin A}{CR \cdot \sin C}$$

$$\Rightarrow \frac{1}{\sin B} = \frac{BQ}{CR \cdot \sin C}$$

$$\Rightarrow CR \cdot \sin C = BQ \cdot \sin B$$

$$\frac{\sin B}{AC} = \frac{\sin C}{AB}$$

$$\Rightarrow \frac{\sin B}{\sin C} = \frac{\frac{CR}{\sin A}}{\frac{AP}{\sin B}}$$

$$\Rightarrow \frac{\sin B}{\sin C} = \frac{CR \cdot \sin B}{AP \cdot \sin A} \quad \Rightarrow \frac{1}{\sin C} = \frac{CR}{AP \cdot \sin A}$$

$$\Rightarrow AP \cdot \sin A = CR \cdot \sin C \quad \Rightarrow AP \cdot \sin A = BQ \cdot \sin B$$

$$\Rightarrow \frac{\sin A}{\frac{1}{AP}} = \frac{\sin B}{\frac{1}{BQ}} = \frac{\sin C}{\frac{1}{CR}}$$

$$\Rightarrow \sin A : \sin B : \sin C = \frac{1}{AP} : \frac{1}{BQ} : \frac{1}{CR}$$

We know $AP : CR : BQ = 3 : 4 : 5$

Let $AP = 3x$, then $CR = 4x$ & $BQ = 5x$

Now

$$\Rightarrow \sin A : \sin B : \sin C = \frac{1}{3x} : \frac{1}{4x} : \frac{1}{5x}$$

$$\Rightarrow \sin A : \sin B : \sin C = \frac{60}{3} : \frac{60}{4} : \frac{60}{5}$$

$$\Rightarrow \boxed{\sin A : \sin B : \sin C = 20 : 15 : 12}$$

Problem 46

In ΔABC

Incircle Radius, $r = 2$ cm

Circumcircle Radius, $R = 3$ cm

$\cos A + \cos B + \cos C = ?$

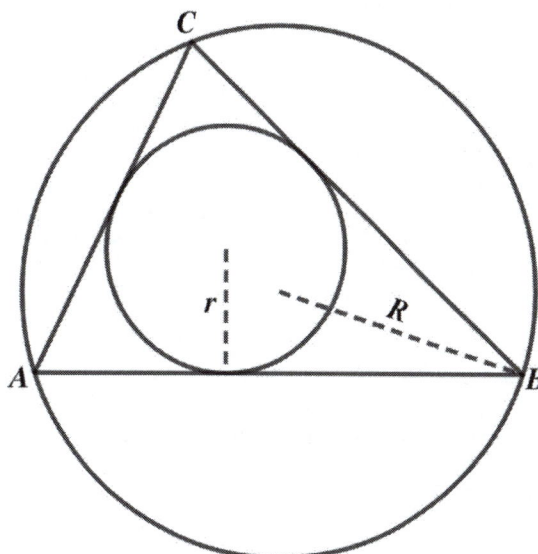

Solution

$$\cos A + \cos B + \cos C = 2 \cdot \cos\left(\frac{A+B}{2}\right) \cdot \cos\left(\frac{A-B}{2}\right) + \cos C$$

$$= 2 \cdot \cos\left(\frac{\pi}{2} - \frac{C}{2}\right) \cdot \cos\left(\frac{A-B}{2}\right) + \cos C \qquad (A + B + C = \pi)$$

$$= 2 \cdot \sin\left(\frac{C}{2}\right) \cdot \cos\left(\frac{A-B}{2}\right) + 1 - 2\sin^2\left(\frac{C}{2}\right) \qquad \left(\cos\theta = 1 - 2\sin^2\left(\frac{\theta}{2}\right)\right)$$

$$= 2 \cdot \sin\left(\frac{C}{2}\right)\left(\cos\left(\frac{A-B}{2}\right) - \cos\left(\frac{A+B}{2}\right)\right) + 1$$

$$= 2 \cdot \sin\left(\frac{C}{2}\right)\left(-2 \cdot \sin\left(\frac{\frac{A-B}{2} + \frac{A+B}{2}}{2}\right) \cdot \sin\left(\frac{\frac{A-B}{2} - \frac{A+B}{2}}{2}\right)\right) + 1$$

$$= 2 \cdot \sin\left(\frac{C}{2}\right)\left(-2 \cdot \sin\left(\frac{A}{2}\right) \cdot \sin\left(\frac{-B}{2}\right)\right) + 1$$

$$= 1 + 4 \cdot \sin\left(\frac{A}{2}\right) \cdot \sin\left(\frac{B}{2}\right) \cdot \sin\left(\frac{C}{2}\right)$$

$$= 1 + 4 \cdot \sqrt{\frac{(s-b)(s-c)}{bc}} \cdot \sqrt{\frac{(s-a)(s-c)}{ac}} \cdot \sqrt{\frac{(s-a)(s-b)}{ab}}$$

$$= 1 + 4 \cdot \frac{(s-a)(s-b)(s-c)}{abc}$$

$$\Rightarrow \quad \cos A + \cos B + \cos C = 1 + 4 \cdot \frac{s(s-a)(s-b)(s-c)}{s \cdot abc}$$

$$= 1 + 4 \cdot \frac{Area^2}{s \cdot abc}$$

$$= 1 + 4 \times \frac{Area}{s} \times \frac{Area}{abc}$$

$$= 1 + 4 \times r \times \frac{1}{4R}$$

$$\Rightarrow \quad \cos A + \cos B + \cos C = 1 + \frac{r}{R}$$

$$\Rightarrow \quad \cos A + \cos B + \cos C = 1 + \frac{2}{3}$$

$$\Rightarrow \quad \cos A + \cos B + \cos C = \frac{5}{3}$$

Problem 47

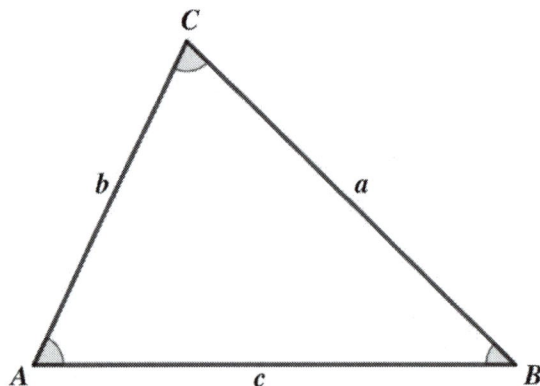

In ΔABC

Prove that

$$\sin A - \sin B + \sin C = 4 \cdot \sin\left(\frac{A}{2}\right) \cdot \cos\left(\frac{B}{2}\right) \cdot \sin\left(\frac{C}{2}\right)$$

Solution

$$2 \cdot \sin\left(\frac{A}{2}\right) \cdot \cos\left(\frac{B}{2}\right) \cdot \sin\left(\frac{C}{2}\right)$$

$$= \left(\sin\left(\frac{A+B}{2}\right) + \sin\left(\frac{A-B}{2}\right)\right) \cdot \sin\left(\frac{C}{2}\right)$$

$$= \sin\left(\frac{A+B}{2}\right) \cdot \sin\left(\frac{C}{2}\right) + \sin\left(\frac{A-B}{2}\right) \cdot \sin\left(\frac{C}{2}\right)$$

$$= \frac{1}{2}\left(\cos\left(\frac{A+B}{2} - \frac{C}{2}\right) - \cos\left(\frac{A+B}{2} + \frac{C}{2}\right)\right)$$

$$+ \frac{1}{2}\left(\cos\left(\frac{A-B}{2} - \frac{C}{2}\right) - \cos\left(\frac{A-B}{2} + \frac{C}{2}\right)\right)$$

$$= \frac{1}{2}\left(\cos\left(\frac{A+B-C}{2}\right) - \cos\left(\frac{A+B+C}{2}\right)\right)$$

$$+ \frac{1}{2}\left(\cos\left(\frac{A-B-C}{2}\right) - \cos\left(\frac{A-B+C}{2}\right)\right)$$

$$= \frac{1}{2}\left(\cos\left(\frac{A+B+C-2C}{2}\right) - \cos\left(\frac{A+B+C}{2}\right)\right)$$

$$+ \frac{1}{2}\left(\cos\left(\frac{2A-(A+B+C)}{2}\right) - \cos\left(\frac{A+B+C-2B}{2}\right)\right)$$

$$= \frac{1}{2}\left(\cos\left(\frac{180-2C}{2}\right) - \cos\left(\frac{180}{2}\right)\right) + \frac{1}{2}\left(\cos\left(\frac{2A-180}{2}\right) - \cos\left(\frac{180-2B}{2}\right)\right)$$

$$= \frac{1}{2}\left(\cos\left(90-C\right) - \cos\left(90\right)\right) + \frac{1}{2}\left(\cos\left(A-90\right) - \cos\left(90-B\right)\right)$$

$$= \frac{1}{2}\left(\sin C - 0\right) + \frac{1}{2}\left(\cos\left(90-A\right) - \cos\left(90-B\right)\right)$$

$$= \frac{\sin C}{2} + \frac{\sin A - \sin B}{2}$$

$$= \frac{\sin C + \sin A - \sin B}{2}$$

$$\Rightarrow\ 2 \cdot \sin\left(\frac{A}{2}\right) \cdot \cos\left(\frac{B}{2}\right) \cdot \sin\left(\frac{C}{2}\right) = \frac{\sin C + \sin A - \sin B}{2}$$

$$\Rightarrow\ 4 \cdot \sin\left(\frac{A}{2}\right) \cdot \cos\left(\frac{B}{2}\right) \cdot \sin\left(\frac{C}{2}\right) = \sin C + \sin A - \sin B$$

$$\Rightarrow\ \boxed{\sin A - \sin B + \sin C = 4 \cdot \sin\left(\frac{A}{2}\right) \cdot \cos\left(\frac{B}{2}\right) \cdot \sin\left(\frac{C}{2}\right)}$$

Problem 48

In $\triangle ABC$, $\sin A + \sin B = \sin\left(\dfrac{A-B}{2}\right)$

Prove that $\dfrac{a-b}{a+b} = 2 \cdot \sin\left(\dfrac{C}{2}\right)$

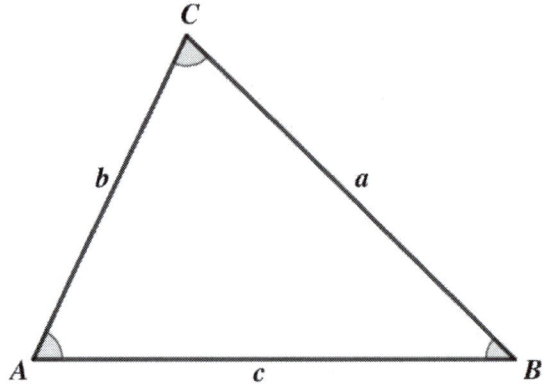

Solution

$$\sin\left(\frac{A-B}{2}\right) = \sin\left(\frac{A}{2}\right) \cdot \cos\left(\frac{B}{2}\right) - \cos\left(\frac{A}{2}\right) \cdot \sin\left(\frac{B}{2}\right)$$

$$\Rightarrow \ \sin\left(\frac{A-B}{2}\right) = \sqrt{\frac{(s-b)(s-c)}{bc}} \cdot \sqrt{\frac{s(s-b)}{ac}} - \sqrt{\frac{s(s-a)}{bc}} \cdot \sqrt{\frac{(s-a)(s-c)}{ac}}$$

$$\Rightarrow \ \sin\left(\frac{A-B}{2}\right) = \sqrt{\frac{s(s-b)^2(s-c)}{abc^2}} - \sqrt{\frac{s(s-a)^2(s-c)}{abc^2}}$$

$$\Rightarrow \ \sin\left(\frac{A-B}{2}\right) = \frac{s-b}{c}\sqrt{\frac{s(s-c)}{ab}} - \frac{s-a}{c}\sqrt{\frac{s(s-c)}{ab}}$$

$$\Rightarrow \ \sin\left(\frac{A-B}{2}\right) = \left(\frac{s-b}{c} - \frac{s-a}{c}\right)\sqrt{\frac{s(s-c)}{ab}}$$

$$\Rightarrow \ \sin\left(\frac{A-B}{2}\right) = \left(\frac{a-b}{c}\right)\sqrt{\frac{s(s-c)}{ab}}$$

$$\Rightarrow \ \sin A + \sin B = \left(\frac{a-b}{c}\right)\sqrt{\frac{s(s-c)}{ab}}$$

$$\Rightarrow \ c \cdot (\sin A + \sin B) = (a-b) \cdot \cos\left(\frac{C}{2}\right)$$

$$\Rightarrow \ c \cdot \sin A + c \cdot \sin B = (a-b) \cdot \cos\left(\frac{C}{2}\right)$$

$$\Rightarrow \quad c \cdot \sin A + c \cdot \sin B = (a - b) \cdot \cos\left(\frac{C}{2}\right)$$

$$\frac{a}{\sin A} = \frac{b}{\sin B} = \frac{c}{\sin C} \qquad (sine\ rule)$$

$$\Rightarrow \quad c \cdot \sin A = a \cdot \sin C$$

$$c \cdot \sin B = b \cdot \sin C$$

Then

$$c \cdot \sin A + c \cdot \sin B = (a - b) \cdot \cos\left(\frac{C}{2}\right)$$

$$\Rightarrow \quad a \cdot \sin C + b \cdot \sin C = (a - b) \cdot \cos\left(\frac{C}{2}\right)$$

$$\Rightarrow \quad (a + b) \cdot \sin C = (a - b) \cdot \cos\left(\frac{C}{2}\right)$$

$$\Rightarrow \quad \frac{a - b}{a + b} = \frac{\sin C}{\cos\left(\frac{C}{2}\right)}$$

$$\Rightarrow \quad \frac{a - b}{a + b} = \frac{2 \cdot \sin\left(\frac{C}{2}\right) \cdot \cos\left(\frac{C}{2}\right)}{\cos\left(\frac{C}{2}\right)}$$

$$\Rightarrow \quad \boxed{\frac{a - b}{a + b} = 2 \cdot \sin\left(\frac{C}{2}\right)}$$

Problem 49

In $\triangle ABC$, $\dfrac{\sin A + \sin B}{\sin A + \sin C} = \dfrac{a+b}{b+c}$

Prove that $\sin A = \cos\left(\dfrac{C}{2}\right)$

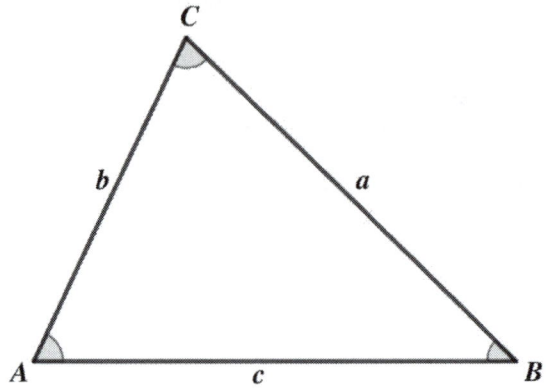

Solution

Let R is the radius of the circumcircle of the $\triangle ABC$

$$\frac{a}{\sin A} = \frac{b}{\sin B} = \frac{c}{\sin C} = 2R \qquad (sine\ rule)$$

$$\Rightarrow \quad \sin A = \frac{a}{2R} \quad \&\ a = 2R\sin A$$

$$\sin B = \frac{b}{2R} \quad \&\ b = 2R\sin B$$

$$\sin C = \frac{c}{2R} \quad \&\ c = 2R\sin C$$

$$\Rightarrow \quad \frac{\sin A + \sin B}{\sin A + \sin C} = \frac{\frac{a}{2R} + \frac{b}{2R}}{\frac{a}{2R} + \frac{c}{2R}}$$

$$\Rightarrow \quad \frac{\sin A + \sin B}{\sin A + \sin C} = \frac{a+b}{a+c}$$

$$\Rightarrow \quad \frac{a+b}{a+c} = \frac{a+b}{b+c} \qquad \Rightarrow\ a+c = b+c$$

$$\Rightarrow\ a = b \quad \Rightarrow\ \angle A = \angle B \quad \Rightarrow\ \angle C = 180 - (\angle A + \angle B)$$

$$\Rightarrow\ C = 180 - 2A \qquad \Rightarrow\ \frac{C}{2} = 90 - A$$

$$\Rightarrow\ \cos\left(\frac{C}{2}\right) = \cos(90 - A)$$

$$\Rightarrow\ \boxed{\cos\left(\frac{C}{2}\right) = \sin A}$$

Problem 50

In ΔABC, Area of the triangle $= 4\sqrt{3}$ cm^2

$\sin\left(\dfrac{A}{2}\right) \cdot \sin\left(\dfrac{B}{2}\right) \cdot \sin\left(\dfrac{C}{2}\right) = \dfrac{1}{8}$ and

$\sin A + \sin B + \sin C = \dfrac{3\sqrt{3}}{2}$

Then, $abc = ?$

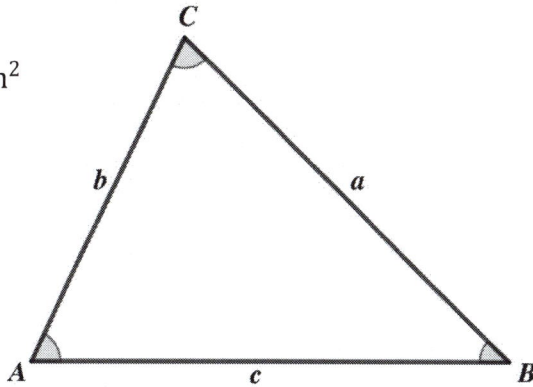

Solution

Let R is the radius of the circumcircle of the ΔABC

$\dfrac{a}{\sin A} = \dfrac{b}{\sin B} = \dfrac{c}{\sin C} = 2R$ \qquad (*sine rule*)

$\Rightarrow \sin A = \dfrac{a}{2R}$ & $a = 2R\sin A$

$\qquad \sin B = \dfrac{b}{2R}$ & $b = 2R\sin B$

$\qquad \sin C = \dfrac{c}{2R}$ & $c = 2R\sin C$

$\sin A + \sin B + \sin C = \dfrac{a}{2R} + \dfrac{b}{2R} + \dfrac{c}{2R}$

$\Rightarrow \sin A + \sin B + \sin C = \dfrac{a+b+c}{2R}$

$\Rightarrow \dfrac{a+b+c}{2R} = \dfrac{3\sqrt{3}}{2}$

$\Rightarrow \dfrac{s}{R} = \dfrac{3\sqrt{3}}{2}$ \qquad $\left(\because s = \dfrac{a+b+c}{2}\right)$

$\Rightarrow s = \dfrac{3R\sqrt{3}}{2}$ .... $eq(1)$

$\sin\left(\dfrac{A}{2}\right) \cdot \sin\left(\dfrac{B}{2}\right) \cdot \sin\left(\dfrac{C}{2}\right) = \sqrt{\dfrac{(s-b)(s-c)}{bc}} \cdot \sqrt{\dfrac{(s-a)(s-c)}{ac}} \cdot \sqrt{\dfrac{(s-a)(s-b)}{ab}}$

$$\Rightarrow \quad \sin\left(\frac{A}{2}\right) \cdot \sin\left(\frac{B}{2}\right) \cdot \sin\left(\frac{C}{2}\right) = \sqrt{\frac{(s-a)^2(s-b)^2(s-c)^2}{(abc)^2}}$$

$$\Rightarrow \quad \frac{1}{8} = \frac{(s-a)(s-b)(s-c)}{abc} \quad \dots\dots\dots\dots\dots\dots\dots\dots\dots\dots\dots\dots\dots\dots eq(2)$$

Area of the triangle, $\Delta = \sqrt{s(s-a)(s-b)(s-c)}$

$$\Rightarrow \quad \Delta^2 = s(s-a)(s-b)(s-c)$$

$$\Rightarrow \quad (s-a)(s-b)(s-c) = \frac{\Delta^2}{s}$$

$$\Delta = \frac{abc}{4R} \quad \Rightarrow \quad abc = \Delta \cdot 4R$$

From $eq(2)$

$$\frac{1}{8} = \frac{(s-a)(s-b)(s-c)}{abc} = \frac{\frac{\Delta^2}{s}}{\Delta \cdot 4R}$$

$$\Rightarrow \quad \frac{1}{8} = \frac{\Delta}{s \cdot 4R} = \frac{4\sqrt{3}}{s \cdot 4R}$$

$$\Rightarrow \quad R = \frac{8\sqrt{3}}{s} \quad \dots\dots\dots\dots\dots\dots\dots\dots\dots\dots\dots\dots\dots\dots\dots\dots\dots\dots eq(3)$$

From $eq(2)$ & $eq(3)$

$$R = \frac{8\sqrt{3}}{s} = \frac{8\sqrt{3}}{\frac{3R\sqrt{3}}{2}}$$

$$\Rightarrow \quad R = \frac{16\sqrt{3}}{R\sqrt{3}} \quad \Rightarrow \quad R^2 = \frac{16}{3} \quad \Rightarrow \quad R = \frac{4}{\sqrt{3}} \text{ cm}$$

$$abc = \Delta \cdot 4R = 4\sqrt{3} \times 4 \times \frac{4}{\sqrt{3}}$$

$$\Rightarrow \quad \boxed{abc = 64}$$

Made in the USA
Monee, IL
08 January 2025